THE INVENTO

ingenious ideas
creative designs

FOOTBALL

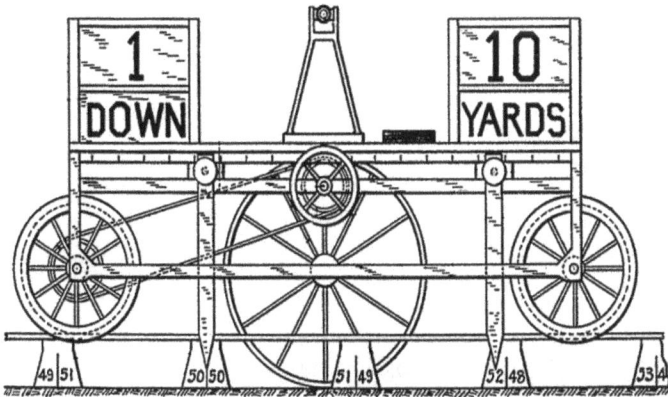

Bad Burro Press
an imprint of Catalyst Publications, Inc.

The Inventors -- creative ideas ingenious designs – FOOTBALL

Source: United States Patent and Trademark Office, www.uspto.gov

Published by:

Bad Burro Press, an imprint of
Catalyst Publications, Inc.
PO Box 2485, Kirkland, WA 98083-2485

catalystpublicationsinc.com

Printed in the United States of America
First Printing 2016

ISBN 978-1-943783-00-7

TABLE OF CONTENTS

Ingenuity

The quality of being clever, original, and inventive.
Derived from the Latin term: ingenuitā, innate virtue or freeborn char-
acteristic.

Each inventor cited in this publication reflects these characteristics.   Regardless
if the individual applications were successfully manufactured or reaped financial
reward, each evidences dedication, determination and creativity.

Patent applications have a long and storied history in the United States.   The
earliest patent was reportedly granted in Massachusetts to Samuel Winslow in
1641 for a process of making salt.   The United States Patent Act of 1790 was
founded upon Article 1, Section 8 of the Constitution signed three years earlier:
"The Congress shall have Power ... To promote the Progress of Science and useful
Arts, by securing for limited Times to Authors and Inventors the exclusive Right
to their respective Writings and Discoveries."

According to the United States Patent and Trademarks Office (USPTO), three
invention patents were granted in 1790 and thirty-three the following year.   The
volume increased: 158 in 1808, 24,656 in 1900, 157,494 in 2000 and 300,678
in 2014.   Creative genius flourishes, and not just within resident citizens.   The
first record of USTPO patents granted to foreign residents was in 1836 with 8.
That number had risen to 167,000 by 2014.

Inventor statements have been excerpted and drawings specially adapted for this
publication.   I hope you enjoy this presentation of ingenuity and creativity.

Christopher Fox

1. FOOTBALL
an ellipsoidal design

One object of the present invention is to provide a football which will be of a balanced character, being substantially no heavier on one side than the other.

I claim as my invention:

In combination with the outer casing of a football having an opening for the insertion of the inflatable member, and a flap to close said opening, an inflatable member having a filling tube to pass through an opening in the flap, and means carried by the opposite side of the ball, and within the same, to balance the weight of said filling tube ...

Inventor: Pierce, George L.
Application Number: US63984223A
Publication Date: 08/19/1924
Filing Date: 05/18/1923

G. L. PIERCE

FOOTBALL

Fig.1.

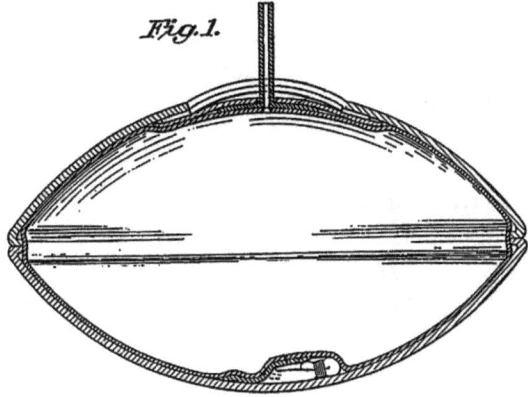

Fig.2.

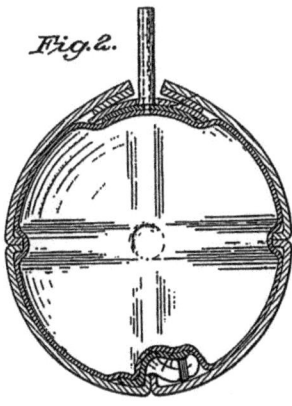

Fig.3.

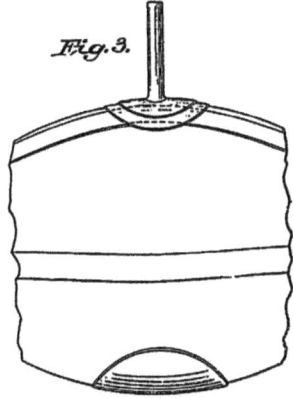

Inventor:

George L. Pierce,

This invention has for its object to provide a football especially adapted for use in playing the present open game wherein the ball is frequently thrown by hand in the play known as the "forward pass". In the execution of this play, it is necessary that the ball be grasped by the end and quickly and accurately thrown, and in wet weather, when the ball is slippery, it is difficult to obtain a sufficiently firm grip thereon to accomplish this successfully. In accordance with the present invention, a football of conventional form is provided at each form with a series of roughened, circumferentially extending lines, preferably formed by stitching, whereby the player is enabled to grip the ball firmly without danger of slipping.

Inventor: Maynard, John E.
Application Number: US66798423A
Publication Date: 10/27/1925
Filing Date: 10/11/1923

J. E. MAYNARD

FOOTBALL

Fig. 1.

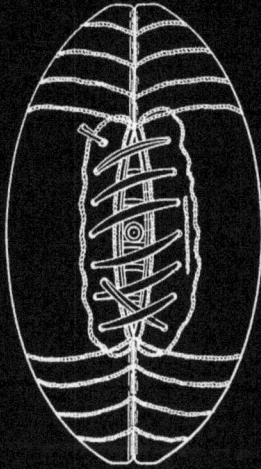

Fig. 2.

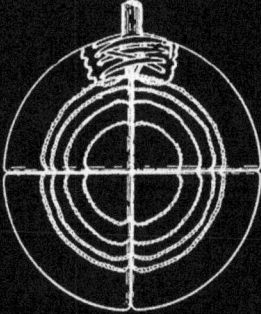

Inventor:

John E. Maynard

7

This invention relates to valve structures for inflatable bladders such-for example as bladders for footballs, basketballs and the like, and is applicable as well to inflatable balls of various types whether or not the inflatable structure is designed to be disposed within a cover. The invention is illustrated, however, in connection with a football, but it will be understood that it is not to be limited to this use.

One object of the invention is the provision of a valve structure for inflatable balls or bladders such that air may be introduced therein with facility and after the operation has been completed the opening through which the air enters may be effectually and easily closed.

Another object of the invention is the provision of an inflatable ball or bladder provided with a seal-sealing inflating structure, which may be manipulated to allow free entry of air therein, and also to seal the opening against the egress of air, and which structure will at the same time be of a simple and economical construction having few parts, and which will be unlikely to get out of order or to need repair.

More specifically the invention resides in providing an inflatable ball of this character with a filler neck extending into the ball and a filler tube or valve tub movably mounted in the neck and of such construction that in one position the entry of air into the ball is permitted, while in another position the ball will be sealed against leakage.

Inventor: Myrick, Crane
Application Number: US53732931A
Publication Date: 11/20/1934
Filing Date: 05/14/1931

Nov. 20, 1934.   M. CRANE

VALVE FOR FOOTBALL BLADDERS AND THE LIKE

Fig.1.

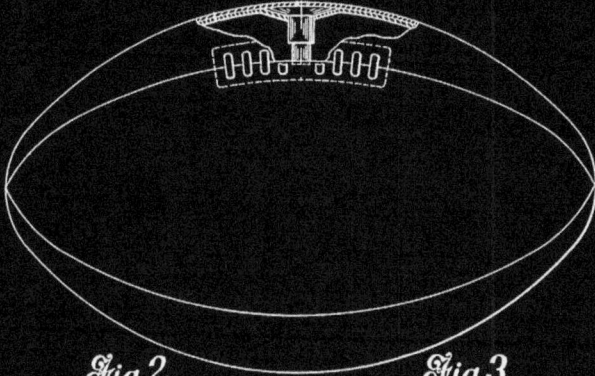

Fig.2.   *Fig.3.*

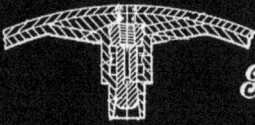   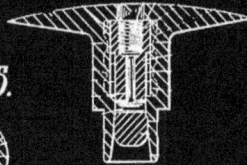

Fig.5.

Fig.4.

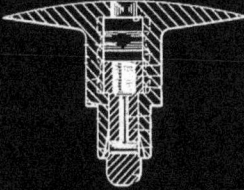   *Fig.6.*   *Fig.7.*

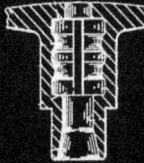

Inventor

Myrick Crane

This invention relates to athletic balls and more particularly to inflatable balls such as footballs of ellipsoidal or generally oval shape.

The invention also relates more particularly to ellipsoidal footballs which are of the so-called carcass type, the wall of the ball being a composite structure including an inflatable valve-equipped bladder and an outer cover of rubber or a plastic having the necessary elasticity.

The invention deals especially with the means employed for the purpose of enabling the ball to be gripped by the hand of the player.

One of the objects of the present invention is to provide a further form of ellipsoidal football which provides a very effective grip upon the ball as it is handled, thrown and caught.

Another object of the invention is to make the gripping of the ball by the player more effective than heretofore so as to reduce fumbling to the minimum.

Another aim of the invention is to provide an ellipsoidal football with a surface confirmation of such character that the ball can be gripped very securely and effectively in the portion near the center of the major axis as well as in the portions adjacent the ends or tips of the ball.

Inventors: Gow, Arthur R. / Madsen, Paul S.
Application Number: US30109252A
Publication Date: 12/30/1958
Filing Date: 07/26/1952

Fig.1.

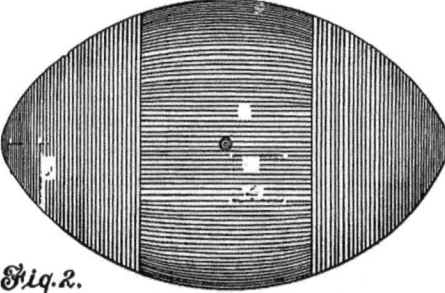

Fig.2.

Fig.4.

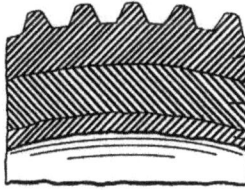

Fig.5.

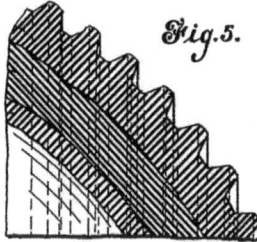

Fig.7.

INVENTORS
a. R. Gow & P. S. Madsen

11

This invention relates to an improvement in footballs and, more particularly, to footballs provided with external fins, preferable integral with the outer cover.

A conventional football, having the shape of a prolate spheroid, requires, for place-kicking, that it be held with its longitudinal axis in a generally vertical position for the kicker. This is normally done by holding the ball by one hand or supporting it on a separate kicking tee, which is usually kicked away when the ball is kicked. The assistance of a teammate, acting as a "spotter," is required if the ball is held by hand; especially in practice kicking, this can be a boring task and provides a source of complaint (or excuse) by the kicker for inaccurate kicks; that is, that the ball was moved or, at least it was not held at the moment of kicking at the angle which the kicker believes proper. If a tee is used, a spotter is not required but time may be lost in placing the ball at the desired angle and in recovering the tee after the kick.

It is the object of this invention to provide a football having fins located at one end so that the ball may be positioned with its axis in a generally verti-cal direction. In addition to the advantage of eliminating the need for the spotter and the loss of time in locating a tee which may be kicked away, an unexpected advantage of this invention is that the fins do not appreciably al-ter the flight of the ball when place-kicked At the same time, the fins appear to make the ball have a truer "spiral" flight when passed or punted; in actual-ity, this latter effect may be more psychological than real, due to the tendency of a passer or punter to try to make a finned ball produced according to this invention take a truer "spiral" flight. …

Thus, a major advantage of this invention is that, in addition to the attractive-ness of the generally rocket-shaped appearance of the ball, it may be used in actual play and behaves substantially the same as conventional footballs and, even when league rules prohibit its use in league play, a football made accord-ing to this invention may be used in practice for improving the players' skill.

Inventor: Smith, Charles E.
Application Number: US27515263A
Publication Date: 06/14/1966
Filing Date: 04/23/1963

C. E. SMITH

FINNED FOOTBALL

Fig.3

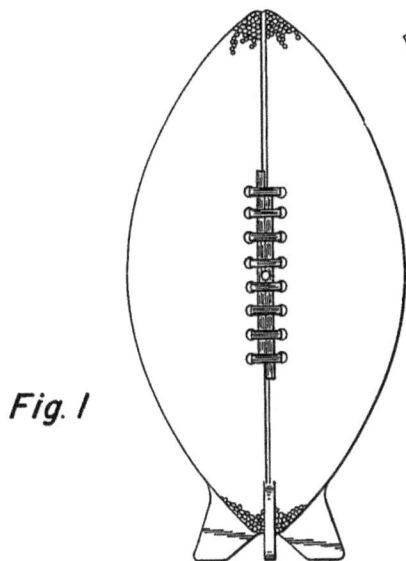

Fig. 1

Fig.4

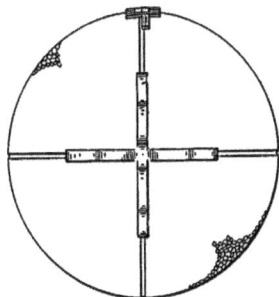

Fig.5

INVENTOR.

CHARLES E. SMITH

Fig. 2

2. HELMET

to protect head and neck

Football helmets have to satisfy widely divergent requirements. Primarily, they are for the protection of the wearer, but in providing this protection care must be exercised lest the means which protect the wearer cause injury to the opponent. Further, a helmet to serve its greatest usefulness must not only yield protection to the top of the head and ears but also to the base of the skull and the neck. Real protection against hard blows can only be secured through an inflexible shell, which can be shaped in manufacture to protect the top of the head and base of the skull and neck. An inflexible shell of this kind, however, may not be placed in contact with head because of the discomfort and injury which would be caused thereby and it may not lie near the outer surface of the helmet because of injury that it might cause to opponents. In view of these considerations, inflexible members, so far as is known, have never been incorporated in football helmets. The principal object of the present invention, therefore, is to meet all the conditions outlined above by providing a football helmet which shall be comfortable to the head, safe for opponents, and afford the greatest possible protection for the wearer through the incorporation therein of an inflexible shell shaped manufactured to protect the head and neck from blows and yet it is to be imbedded in protective coverings of felt or the like. A further object of the invention is to provide in such a helmet a smooth continuous outer covering therefor which shall prevent injury to the helmet and present a comparatively smooth surface to opponents.

Inventor: Ridgeway, Hart Henry
Application Number: US52843222A
Publication Date: 02/19/1924
Filing Date: 01/11/1922

Feb 19 1924

H. R. HART

FOOTBALL HELMET

Fig.1.

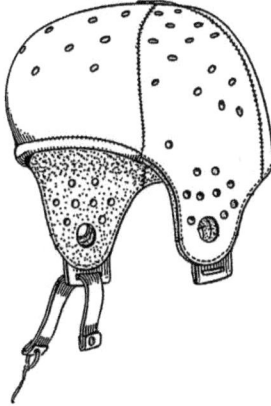

Fig.2.

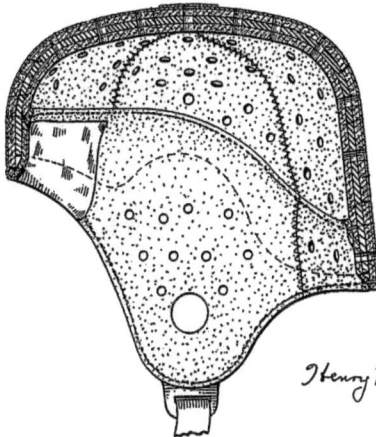

INVENTOR
Henry Ridgeway Hart

An object of my invention is to provide a foot ball helmet having certain new protective features.

Another object of my invention is to provide a device of the class referred to, wherein the parts are so related that the helmet will retain its position upon the head of the wearer without the use of chin straps or the like.

Another object of my invention is to provide a helmet that will fully protect the base of the skull.

Inventor: Hugo, Goldsmith
Application Number: US60215222A
Publication Date: 01/13/1925
Filing Date: 11/20/1922

H. GOLDSMITH

FOOTBALL HELMET

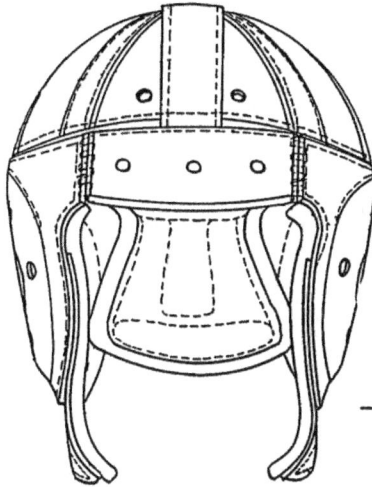

FIG 1

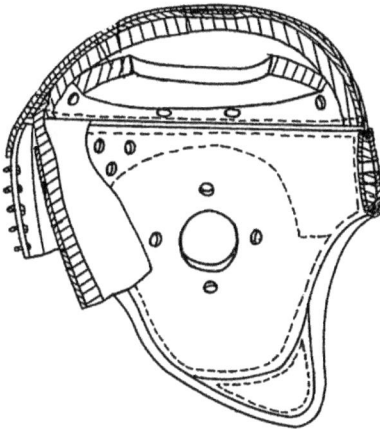

FIG

Inventor
GOLDSMITH

This invention relates to improvements in football helmets, and it consists of the constructions, combinations and arrangements herein described and claimed.

One of the important objects of the invention is to construct the ear guards or protectors in such a manner as to transfer the force of the blow to the skull of the wearer, the arrangement providing a recess or pocket by which the ear of the wearer is received in a partially folded over position, such recess being sufficiently deep to fully contain the ear and to make it impossible for any part of a blow to reach the ear, thus preventing what is commonly known as "cauliflower ears".

Inventor: Kennedy, Allen E.
Application Number: US18248027A
Publication Date: 11/08/1927
Filing Date: 04/09/1927

Nov. 8, 1927.

A. E. KENNEDY

FOOTBALL HELMET

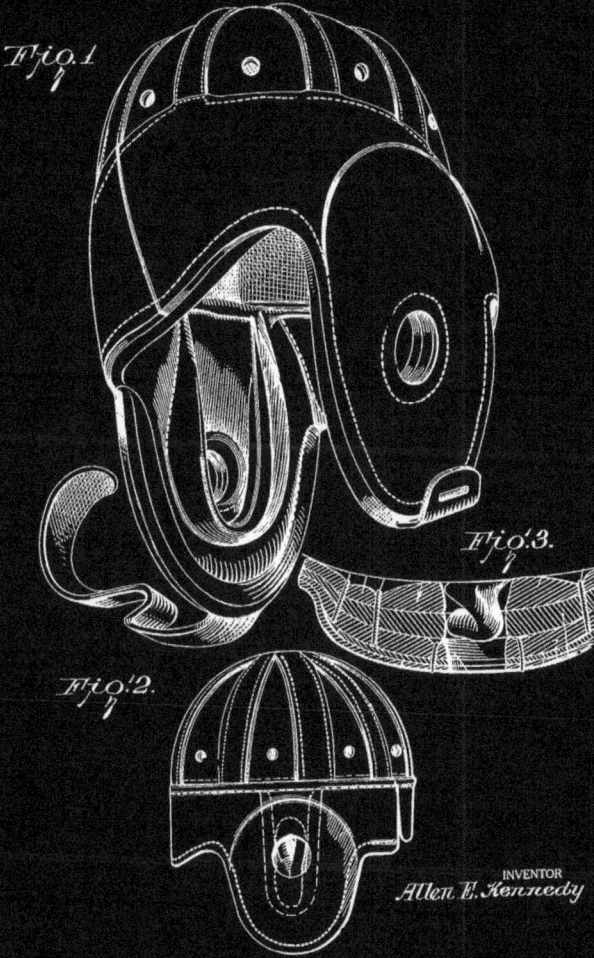

Fig.1

Fig.3.

Fig.2.

INVENTOR

Allen E. Kennedy

The present invention relates to helmets, particularly those adapted for use in playing football, and has for an object the provision of a helmet, the dome of which is constructed in such a manner that it will protect the head of the player against shocks resulting from excessive blows. ...

An object of the present invention is to provide a helmet constructed along different lines wherein leather of a sufficiently rigid character is used and formed into a helmet. An object of this invention is to provide a helmet of the above character to protect the head from the most severe blows received thereby while having in mind that it must not present an outer surface of too hard a nature so that injury to other players might result.

Inventor: Mullins, Robert T.
pplication Number: US17386527A
Publication Date: 01/15/1929
Filing Date: 03/09/1927

R. T. MULLINS

FOOTBALL HELMET

Fig. 1.

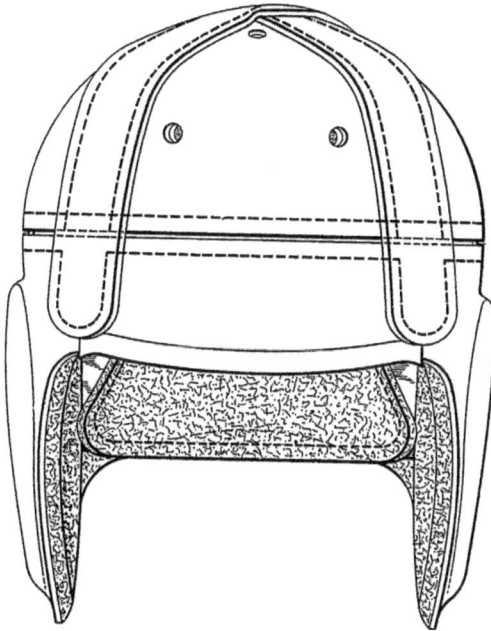

INVENTOR
Robert T. Mullins

My invention relates to protective helmets, especially of the type used by football players, and concerns improvements in the construction of the helmet at the region of the forehead.

Among the objects of my improved construction are: a salient bumper to help keep objects from hitting the wearer's face; semi rigidity for the region of the forehead; distributing the strain of an impact at the forehead over a relatively large area; additional thickness of padding across the forehead without sacrifice of a neat appearing stream line exterior; a convenient finger grip for quacking pulling the helmet on or off the head; and prevention of unbuckling of the forehead strap under strain of the inner crown suspension strap.

Inventor: Turner, Archibald J.
pplication Number: US42862930A
Publication Date: 01/26/1932
Filing Date: 02/15/1930

Jan. 26, 1932.   A. J. TURNER

FOOTBALL HELMET

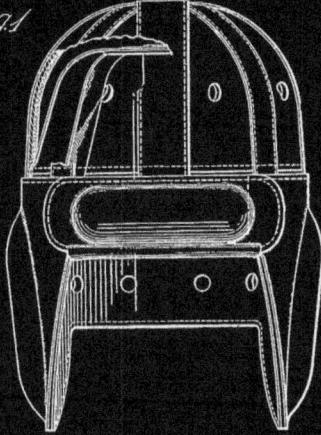

Fig.1

Fig.2

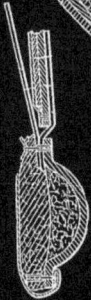

Fig.3

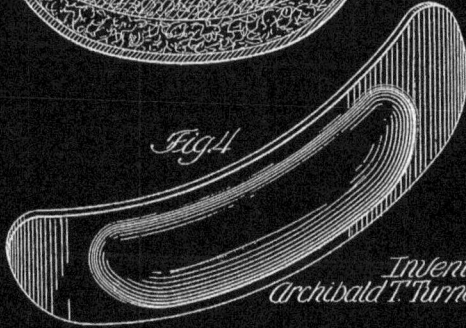

Fig.4

Inventor
Archibald T. Turner

25

This invention relates to body protectors of the type employed for protecting participants in violent games, such as football.

The object of the invention is to prevent the injurious effects to the body of either an extremely localized force from a blow or sudden application to a part of the body itself of the entire force of a blow, although the same may be distributed over a large area. Stated differently, the principal object of the invention is to provide first for a shock absorbing local yielding in respnse to a blow, then the distribution of the force of the blow over a localized area of substantial extent, and then for the shock absorption of the said locally distributed force.

In carrying out my invention, I have found it particularly applicable to the protection of the head of football players and have found that a satisfactory embodiment may be fabricated functionally out of three parts; an outer layer relatively thin, of resilient elastic material such foam or sponge rubber, a thicker inner layer of the same foam or sponge rubber, and intermediate the two layers, ridges preferably discontinuous and out of contact with one another of strong, stiff material such a cane or wood.

Inventor: Taylor, James P.
Application Number: US64059532A
Publication Date: 04/09/1935
Filing Date: 11/01/1932

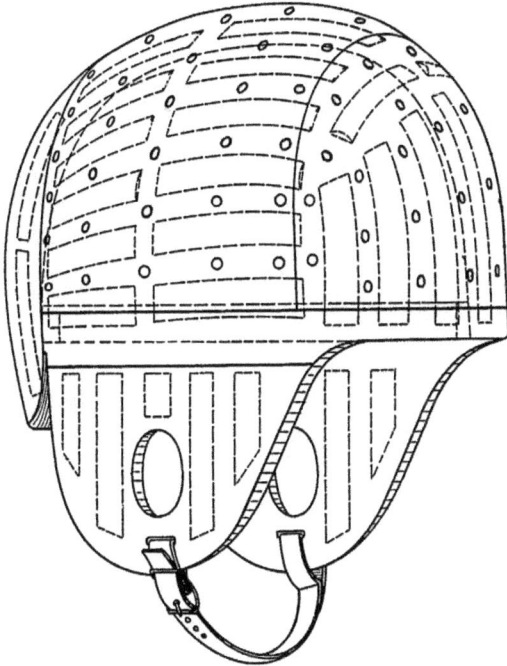

INVENTOR
JAMES P. TAYLOR

This invention relates to football helmet, one of the objects being to provide an inflatable helmet for cushioning any blows received thereby so that the force of the blow will be distributed evenly to all portions of the inner surface of the helmet and injury to the wearer thus reduced to the minimum. A further object is to provide a helmet which, when not in use, can be deflated and thus compactly stored. A further object is to provide a helmet the exposed surfaces of which are formed of rubber so that they will not absorb moisture and acquire objectionable odors. A still further object is to provide an inflatable helmet having new and improved means for holding it in proper shape when inflated and for maintaining desired ventilation while being worn.

Inventor: Harvey Holstein
Application Number: US26019439A
Publication Date: 03/26/1940
Filing Date: 03/06/1939

March 26, 1940.

H. HOLSTEIN

FOOTBALL HELMET

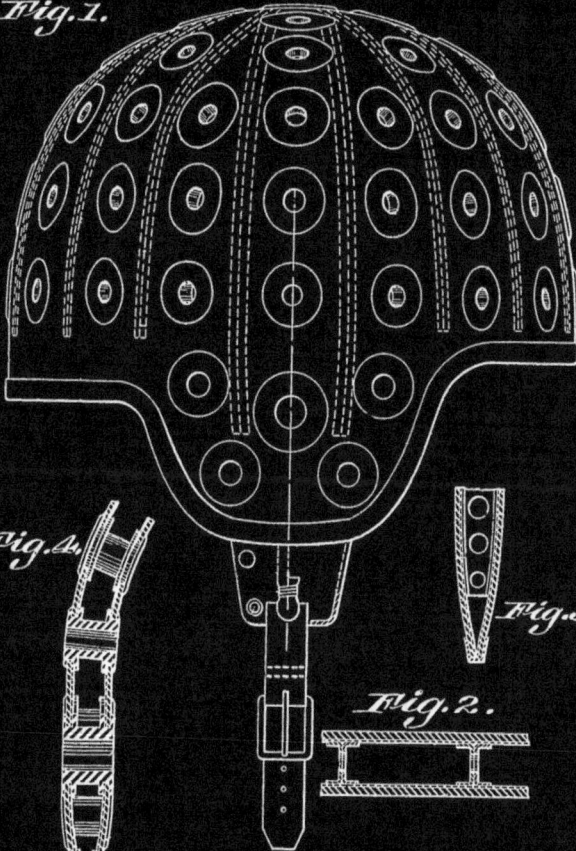

Fig.1.

Fig.4.

Fig.3.

Fig.2.

Harvey Holstein
INVENTOR.

This invention relates to a protective helmet, and particularly to one formed from a unitary skull shaped plate of hard resilient waterproof material.

There has been for some time a considerable demand for a protective helmet of resilient waterproof material of great strength   Helmets of this type are necessary in football, polo, and certain other strenuous sports, and also in many occupations, such, for example, as for the crew of a tank, racing drivers, army aviators, or in any other occupation where the head is likely to be brought into violent contact with any hard object.

The helmets now in use are either leather or similar pads or are built up from a combination of fibrous material or fibre board in combination with other parts such as lining, cover, etc.

All of the materials now employed for this purpose are extremely susceptible to water, either from outside the helmet or water generated by perspiration inside the helmet, and once wet, lose their effectiveness as well as becoming heavy, cold and clammy. At the same time none of the protective devices now on the market having any strength can be made in a single piece, and none of them have the shock, impact or wear resistance desirable in a helmet.

In accordance with the present invention a helmet is prepared which is completely waterproof, which may be prepared in a single piece, which has tensile strength, impact resistance and wear resistance far exceeding anything heretofore on the market, but which is still resilient.

Inventor: Clark, John T.
Application Number: US34565640A
Publication Date: 10/06/1942
Filing Date: 07/15/1940

J. T. CLARK

PROTECTIVE HELMET

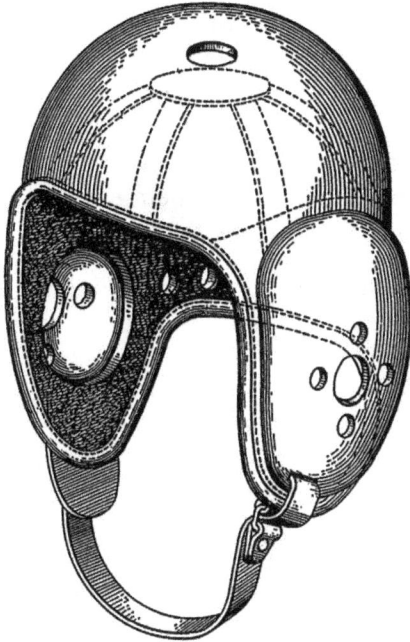

Inventor:
John T. Clark

31

This invention relates to a novel body harness adapted to be connected to the football helmet of a wearer, for preventing neck and head injuries.

The primary object of the invention is the provision of an efficient, practical, and easily used harness of the kind indicated, which includes a body-encircling member, to be worn around the body below the shoulders, and a connecting member extending upwardly from the body encircling member to the forepart of the helmet, and adapted to be connected thereto, preferably to the faceguard bars of the helmet, where so equipped, so that in body contacts during the course of a football game, the head of the player is prevented from being injuriously pushed rearwardly, relative to his shoulders, and the helmet is prevented from being pushed rearwardly, on the head of the player, so as to prevent exposure of the fact of the player to injury, and/or prevent injurious contact of the rear edge of the helmet with the back of the head or neck of the player.

Another object of the invention is the provision of a device of the character indicated above, which is easily and quickly applicable to and removable from the body and helmet of a player, and which, while desirably limiting rearward binding of the head of the player, does not interfere with the player's normal activities, and serves to maintain the player's head in line with his spine, in a position to non-injuriously resist contacts with or blows upon the top or side of the head.

Inventor: Jones, Cordell C.
Application Number: US17461562A
Publication Date: 09/15/1964
Filing Date: 02/20/1962

FIG. 1.

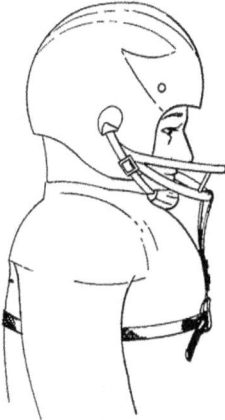

FIG. 3.

FIG. 2.

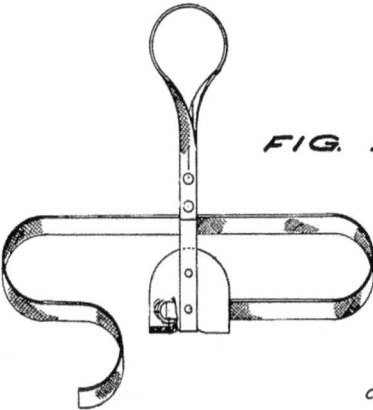

INVENTOR.
CORDELL C. JONES,

This invention relates to head protectors and more particularly to football or crash helmets.

As is generally known, crash helmets at the present time are adaptions of football helmets. However, the evolution of such headgear started during ancient times when warriors required head protection from rocks and stones that were hurled at them. These ancient head protectors were heavy leather hats which fitted snug on a warrior's head to prevent a skull form splitting and being pierced. Early football head gear was similarly made and ear flaps were added.

Later modifications included padding, a back drape between the ear flaps to protect the base of the skull, and then a resilient or elastic sling between the padded leather helmet and head of the wearer. The sling does not necessarily contour to the particular shape of the wearer's head. With the advent of plastics, leather was replaced. Plastic cases were more reasonable to manufacture, stronger and can be permanently colored. However, plastic helmets are known to break laying the wearer open to injury. Thus, plastic helmets did not alleviate all head injuries suffered by wearers of leather helmets which were weaker than those of plastic.

Accordingly, an object of this invention is to provide a football type helmet of the plastic variety which remains intact after a severe shock.

Another object of this invention is to provide a football type helmet which is a resiliently connected case portion which resists breakage when struck.

Another object of this invention is to provide a strong and rigid head protector which resists breaking when struck and has resilient means for absorbing the force of impact.

Another object of this invention is to provide snugness and adaptation to fit to the contour of the wearer's head and yet maintain space between the head and the outer case.

Another objection of this invention is to provide a means for release of the face guard when it is grabbed or pulled to avoid resultant whiplash of the neck caused when the face guard and helmet are integral.

Inventor: Zygmund, Nedwick
Application Number: US17626062A
Publication Date: 10/20/1964
Filing Date: 02/28/1962

Z. NEDWICK

ROTARY FOOTBALL HELMET

Fig. 1

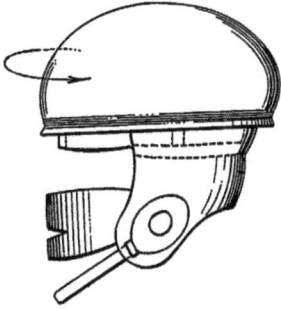

Fig. 3

Fig. 2

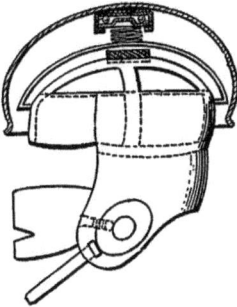

INVENTOR

ZYGMUND NEDWICK

3. GEAR
prepare to play

This invention relates to improvements in goggles designed to mount unbreakable prescription lenses and otherwise stand rough usage, so as to be adapted for use in sports such as football, hockey, and soccer by athletes whose defective vision would not permit their participation in these sports without such air.

The principal object of my invention is to provide goggles having sufficient rigidity to hold the unbreakable lenses securely in proper relation to the eyes while having sufficient flexibility and cushioning action to withstand the shocks and contacts incident to a sport such as football, the goggles being further designed so that the hazard of scratching the lenses is reduced to a minimum.

Another important object of the invention is to provide a goggle frame of one size which upon insertion of the proper sized lenses will fit practically any potential user within a wide range regardless of differences in pupillary measurement. This enables quick, accurate and economical fitting, for all the information that is required with an order for goggles for a certain player is the pupillary distance and the prescribed correction for each eye.

Inventor: Ryan, Donald T.
Application Number: US31287740A
Publication Date: 06/08/1943
Filing Date: 01/08/1940

June 8, 1943.   D. T. RYAN

FOOTBALL GOGGLES

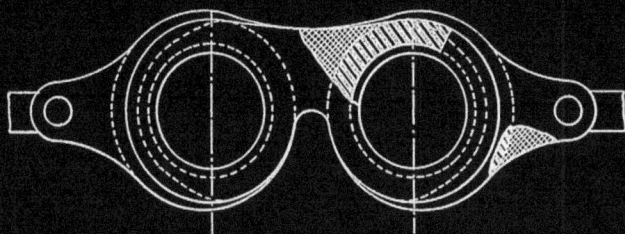

Fig. 1

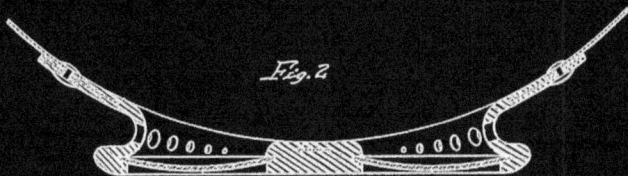

Fig. 2

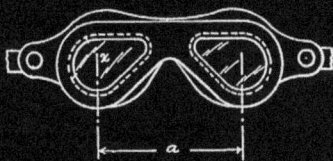

Fig. 3

Fig. 5

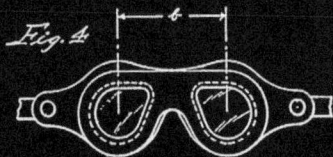

Fig. 4

Fig. 6

Inventor:
Donald T. Ryan

This invention relates to an improved football jersey having hand-hold means.

The primary object of the invention is to provide hand-hold means on football jerseys arranged to be grasped by the hands of offensive blockers while blocking so that neither of the hands is free to deliberately or inadvertently and illegally grasp and hold defensive players, so that holding penalties are prevented.

Another object of the invention is to provide hold-hand means of the character indicated above, which are preferably in the form of strap, and which are detachably secured to a foot-ball jersey so that they will become detached from the jersey, if grasped and pulled by a tackler, so that they cannot be availed of as handles in tackling or in holding the wearer of the jersey.

A further object of the invention is the provision of hand-hold straps of the character indicated above which can be made in rugged and serviceable forms at relatively low cost, which do not add objectionably to the overall cost of football jerseys, which do not interfere with the laundering of the jersey, and which present no inconvenience or burden to a football player.

Inventor: Corbett, Donald T.
Application Number: US76363158A
Publication Date: 12/26/1961
Filing Date: 09/26/1958

D. T. CORBETT

FOOTBALL JERSEY HAVING HAND-HOLDS

FIG.1

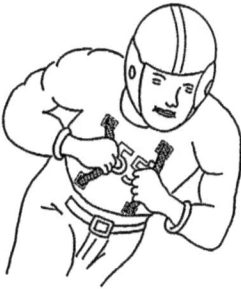

FIG.2

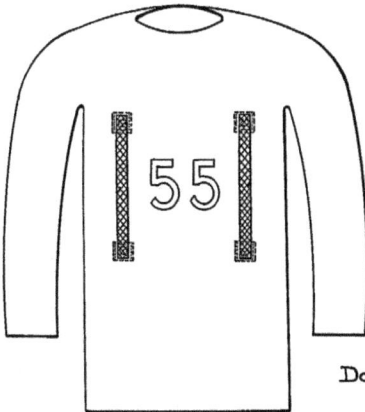

INVENTOR.
DONALD T. CORBETT

In my invention, I provide a football glove having a wrist strap, a palm portion and finger and thumb stalls with the back of the glove being open. The backs of the finer stalls extend only from the ends of the fingers to between the first and second joints thereof. I provide patches of a tenacious gripping material on the tip portions of said thumb and finger stalls. I preferably provide elastic tension bands across the back of the finger stalls between the first and second joints to secure the wearer's fingers in the finger stalls. I also preferably provide an elastic strap attached at its ends to the palm portion of the glove and extending across the back of the wearer's hand in a zone removed from the joins and knuckles of the wearer's hand. I further preferably provide additional patches of tenacious gripping material disposed in the pad zones of the palm of the glove.

Thus, I have invented a glove which provides not only a tenacious grip for all ball handlers but also provides for the freedom, flexibility and maneuverability of the hand and in particular the fingers so necessary for competent manipulation of the ball.

Another advantage of my glove is that to some extent it will protect and warm the hand of the wearer which certainly in foul weather will be of benefit to the wearer in keeping his hand more manipulative than it ordinarily would be in such circumstances.

Inventor: Bruchas, George R.
Application Number: US15105661A
Publication Date: 07/09/1963
Filing Date: 11/08/1961

July 9, 1963   G. R. BRUCHAS

FOOTBALL GLOVE

Fig. I.

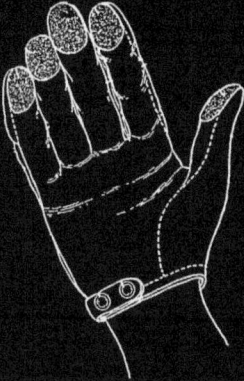

Fig. 2.

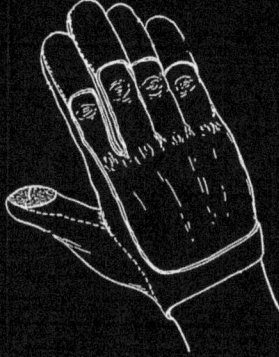

Fig. 3.

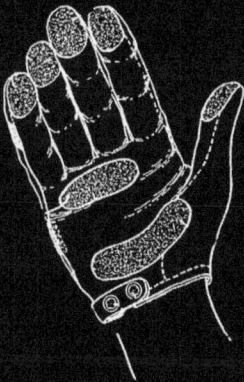

Fig. 4.

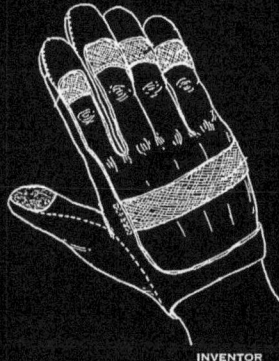

INVENTOR

GEORGE R. BRUCHAS

43

This present invention relates to improvements in a ball player's mask affording the wearer better protection of his face, or particularly of his note.

The principal object of the invention is to provide in a mask of plastic exterior layer an interior one-piece padding with a single cut-out for the nose of the player, the mouth hole and eye holes of the outer layer.

Inventor: Marietta, Michael T.
Application Number: US13577749A
Publication Date: 12/26/1950
Filing Date: 12/29/1949

Dec. 26, 1950

M. T. MARIETTA

PLASTIC FOOTBALL PLAYER'S MASK

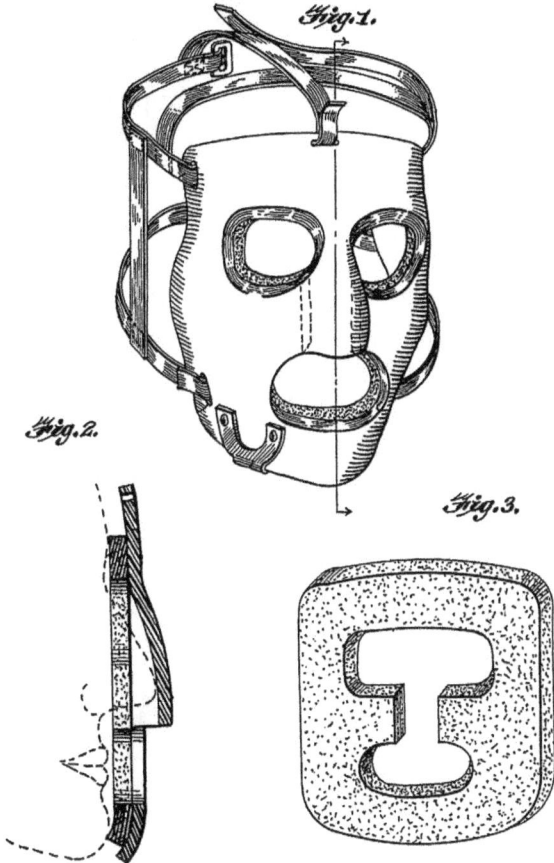

Fig.1.

Fig.2.

Fig.3.

INVENTOR.
MICHAEL T. MARIETTA

The present invention relates generally to protective apparatus, and more particularly to improved equipment for protecting the head and neck of a football player.

The recent increase in head and neck injuries occurring to football players and the seriousness of these injuries has resulted in considerable effort being directed to providing equipment which offers greater protection to the head and neck of a football player against serious injuries. Changes have been suggested in the form of the football helmet and efforts made to eliminate the face guard presently attached to the football helmet. However, the proposed changes provide only a minor reduction in the amount of force applied to the head and neck of a football player and would not greatly reduce or substantially eliminate the danger of severe head and neck injuries.

It is therefore an object of the present invention to provide improved protective equipment for football players which will substantially reduce serious injuries to the head and neck.

Inventors: Archie, Shaffer / Irwin, Shaffer
Application Number: US19149662A
Publication Date: 05/26/1964
Filing Date: 05/01/1962

A. SHAFFER ETAL

PROTECTIVE FOOTBALL APPARATUS

Fig.1

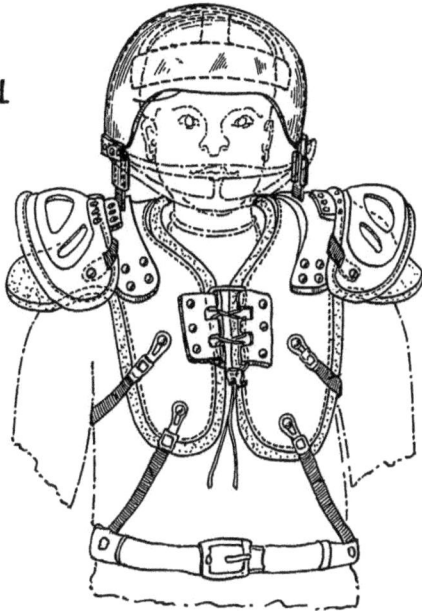

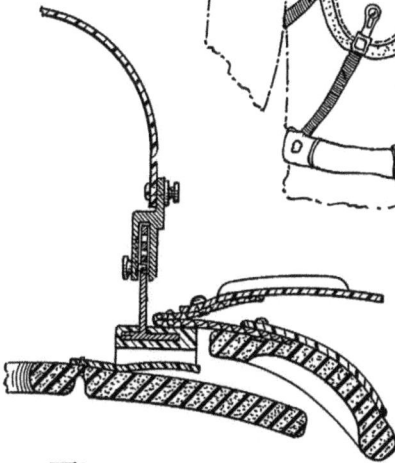

Fig.3

Inventors
Archie Shaffer
Irwin Shaffer

47

4. PRACTICE
the fundamentals

My invention relates to football training devices and more particularly is directed to devices for developing accuracy and confidence in kicking a football.

The general object of my invention is to develop skill in place kicking and to do so without the necessity of identifying, analyzing, and correcting the defects in a kicker's form. My invention is characterized by the concept of confining the player to an accurate technique in practice kicks. I propose to provide physical guide means whereby a player with little or no supervision may repeat correct and accurate practice kicks until correct performance and accurate placing becomes so habitual as to make the guide means unnecessary. One object of my invention is to provide a straight guide channel that is wide enough to permit the forward movement therethrough of a player's foot in approaching and kicking the ball, but is narrow enough to restrict such forward path of the foot substantially to a straight line. Another object is to provide such a channel of approach in combination with means for positioning a football in alignment with the axis of the channel. In the preferred form of my invention, one object is to guide the forward movement and final positioning of the foot on which the player stands when delivering the kick.

A still further object of the invention is to provide means, preferably in the form of an attachment, for guiding a player's foot throughout the arc of a kicking stroke, whereby the player may practice the kicking stroke with repeated accuracy.

Inventor: Peterson, Millard W.
Application Number: US31282840A
Publication Date: 04/22/1941
Filing Date: 01/08/1940

M. W. PETERSON

DEVICE FOR TRAINING FOOTBALL KICKERS

Fig.3.

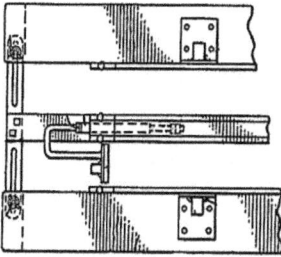

Fig.1.

Fig.4.

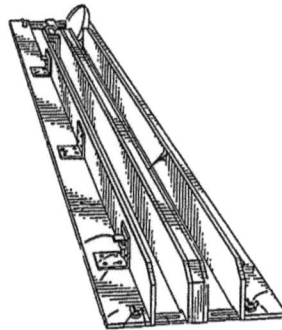

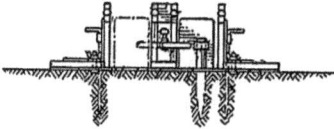

Fig.5.

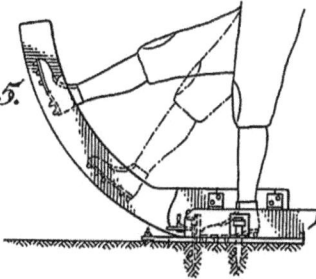

Fig.2.

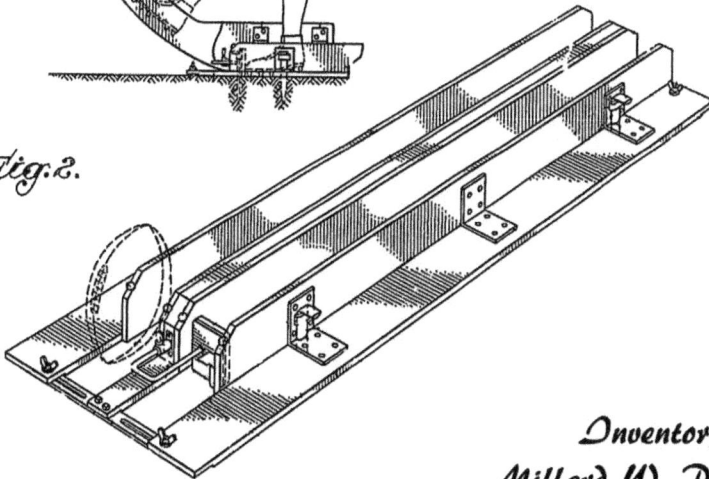

Inventor,
Millard W. Peterson

The main object of my invention is to provide a device or a means for promoting proper lifting action of the legs in running and for use in general in conditioning a runner through special exercise.

An ancillary object of the invention is to provide means for encouraging exercise of the legs of such character that the runner is bound to step over regularly spaced obstructions while running.

It is also an object to have such a runway that has yielding parts to minimize danger of injury to a runner in cases of tripping for falling or becoming confused while running.

An important object of this invention is to provide means for guiding and determining the length of the steps taken by a runner, indoors or outdoors, that can be set up or taken down quickly and conveniently, and which may be stored away when not in active use.

Inventor: Zygmund, Nedwick
Application Number: US35576153A
Publication Date: 08/03/1954
Filing Date: 05/18/1953

Aug. 3, 1954

Z. NEDWICK

GUIDE RIB RUNWAY

Fig. 1

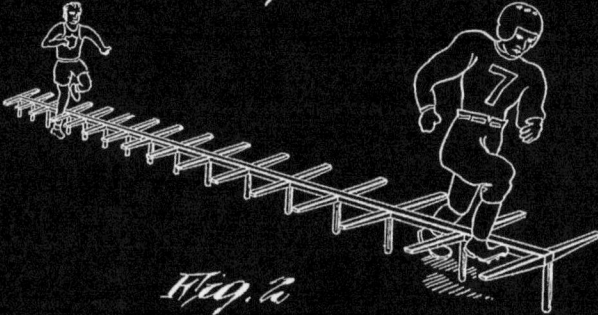

Fig. 2

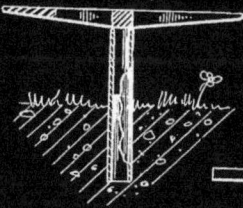

Fig. 3

Fig. 4

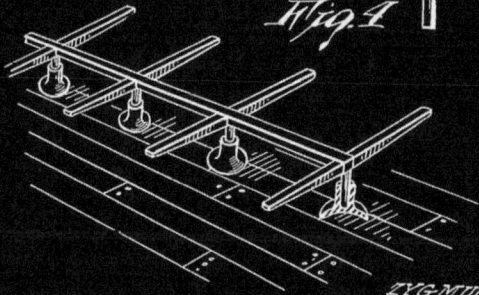

INVENTOR.

ZYGMUND NEDWICK

53

This invention relates to a sound mind in a sound body and particularly to the physical training of youth and somewhat older persons in the kicking of an object such as a football, soccer ball, or the like, and it is an object of the invention to teach or instruct in the necessary fundaments including the relative positions of the foot and leg during the kicking operation and the relative straight locked position of the ankle so that upon impact of the foot with the football the entire foot will be relatively rigid resulting in greater accuracy and distance as, for example, when kicking a field goal.

Inventors: Lee, Ryals E. / Houston, James E. / Norman, George M.
Application Number: US56179666A
Publication Date: 06/27/1967
Filing Date: 06/30/1966

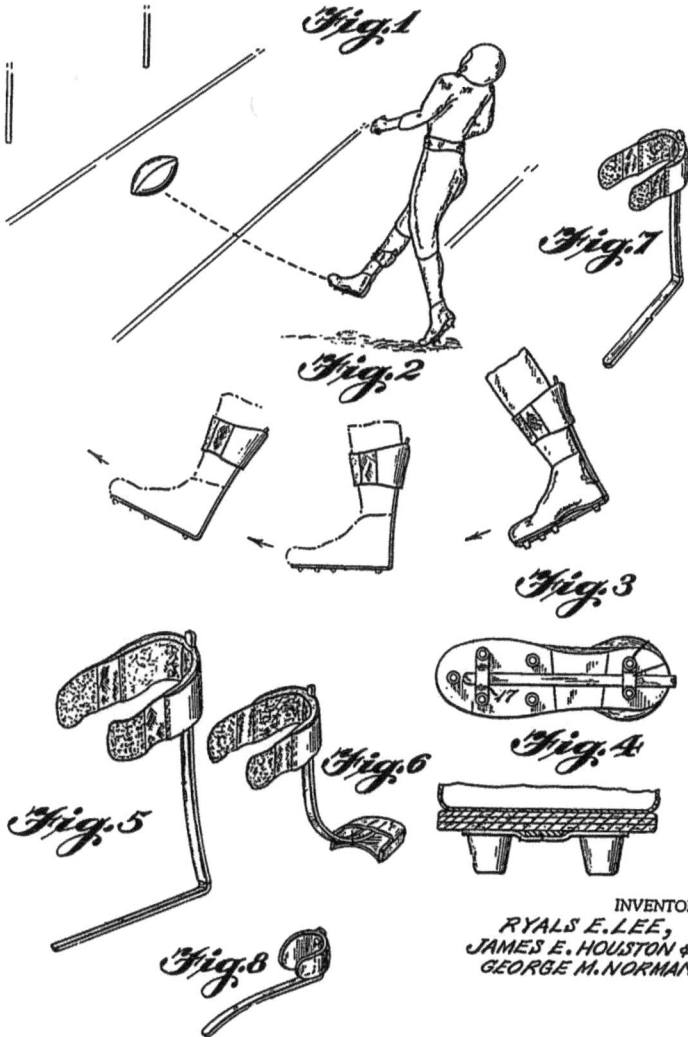

Fig.1

Fig.7

Fig.2

Fig.3

Fig.5

Fig.6

Fig.4

Fig.8

INVENTORS
RYALS E. LEE,
JAMES E. HOUSTON &
GEORGE M. NORMAN

55

5. KICKING
objective and objects

This invention relates to the improvements in football tees and has for its primary object the provision of a device for supporting a football off the ground in a position to be kicked.

Another object of the invention resides in a football tee which is adapted to be used in the playing of the game of football to support the football at the proper angle to be kicked whereby the necessity of having one of the players hold the football in such position is dispensed with. It will therefore be appreciated that the team kicking off will present an equal number of men in action against the team receiving the kick-off, as theretofore, the player assigned to holding the ball was at a decided disadvantage in getting into action after the ball was kicked.

A further object is to provide a football tee which is constructed of a single piece of material bent to form a pair of yieldable jaws for gripping the football, and a shank adapted to be inserted into the ground for supporting the tee in an upright position.

Inventor: Martin, Boettcher
Application Number: US13473326A
Publication Date: 11/23/1926
Filing Date: 09/10/1926

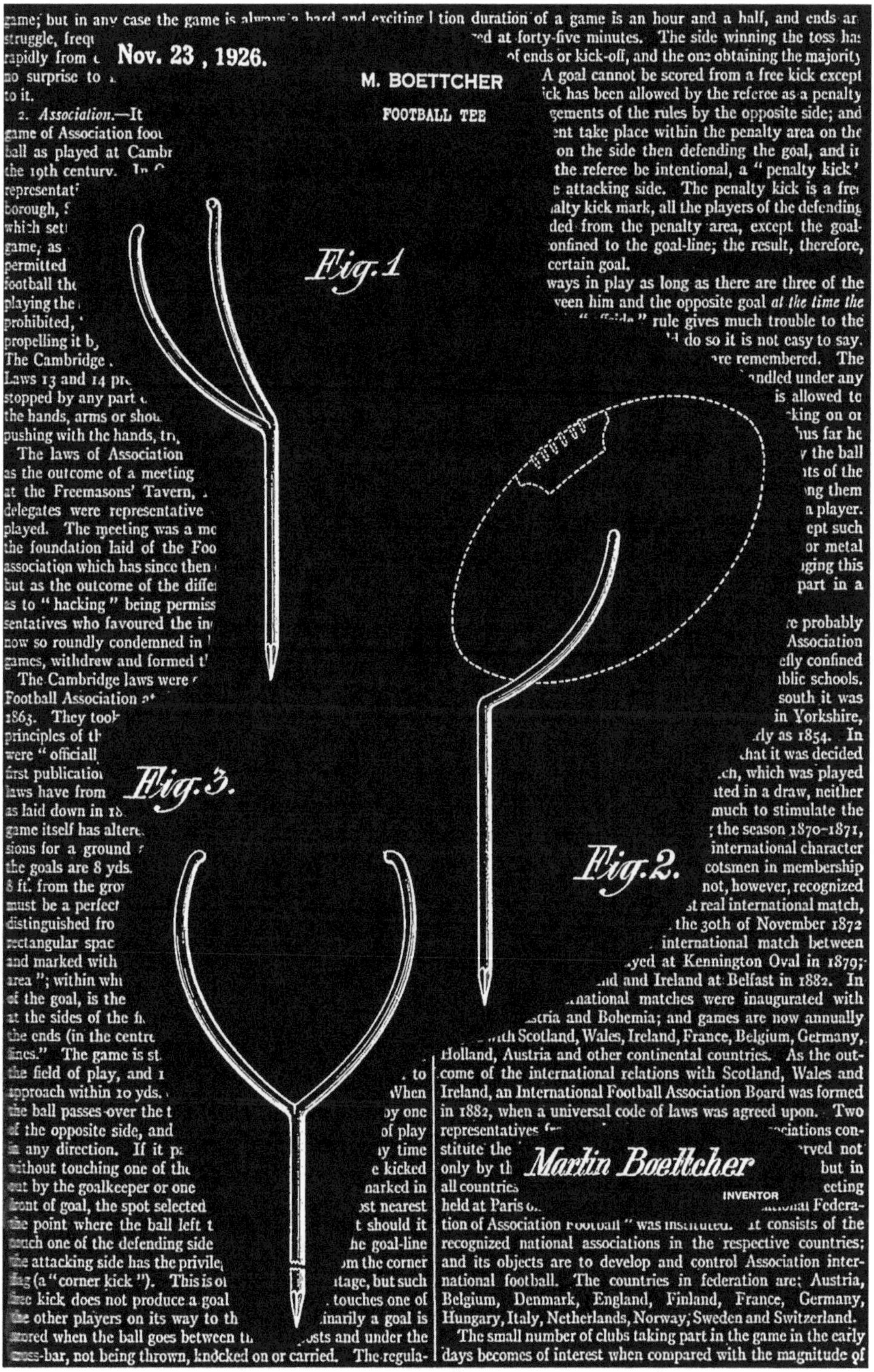

game; but in any case the game is always a hard and exciting
struggle, freq...
rapidly from c...
no surprise to ...
to it.

2. *Association.*—It
game of Association foot
ball as played at Cambr
the 19th century. In C
representati...
borough, S...
which set...
game, as
permitted
football the
playing the
prohibited, ...
propelling it b,
The Cambridge .
Laws 13 and 14 pro
stopped by any part c
the hands, arms or shou
pushing with the hands, tr,
The laws of Association
as the outcome of a meeting
at the Freemasons' Tavern, ...
delegates were representative
played. The meeting was a me
the foundation laid of the Foo
association which has since then
but as the outcome of the differ
as to "hacking" being permiss
sentatives who favoured the in
now so roundly condemned in
games, withdrew and formed t'
The Cambridge laws were ...
Football Association a+
1863. They took
principles of th
were "officiall
first publication
laws have from
as laid down in 18
game itself has altere...
sions for a ground ...
the goals are 8 yds.
8 ft. from the gro...
must be a perfect
distinguished fro
rectangular spac
and marked with
area "; within wh...
of the goal, is the
at the sides of the fi...
the ends (in the centre
lines." The game is st
the field of play, and ...
approach within 10 yds. ...
the ball passes over the t
of the opposite side, and
in any direction. If it p...
without touching one of the
out by the goalkeeper or one
front of goal, the spot selected
the point where the ball left t
touch one of the defending side
the attacking side has the privile
kick (a "corner kick"). This is o...
kick does not produce a goal
the other players on its way to th
scored when the ball goes between th
cross-bar, not being thrown, knocked on or carried. The regula-

M. BOETTCHER

FOOTBALL TEE

Fig.1

Fig.3.

Fig.2.

tion duration of a game is an hour and a half, and ends ar
...ed at forty-five minutes. The side winning the toss ha...
...of ends or kick-off, and the one obtaining the majority
A goal cannot be scored from a free kick except
...ick has been allowed by the referee as a penalty
...gements of the rules by the opposite side; and
...ent take place within the penalty area on the
...on the side then defending the goal, and ir
the referee be intentional, a "penalty kick'
...e attacking side. The penalty kick is a fre
...alty kick mark, all the players of the defending
...ded from the penalty area, except the goal-
...onfined to the goal-line; the result, therefore,
certain goal.
...ways in play as long as there are three of the
...veen him and the opposite goal *at the time the*
..."...i..." rule gives much trouble to the
...d do so it is not easy to say.
...are remembered. The
...ndled under any
...is allowed to
...king on or
...us far he
...y the ball
...ts of the
...ng them
...a player.
...ept such
...or metal
...ging this
...part in a

...re probably
...Association
...efly confined
...iblic schools.
...south it was
...in Yorkshire,
...dy as 1854. In
...hat it was decided
...ch, which was played
...ited in a draw, neither
...much to stimulate the
...; the season 1870–1871,
...international character
...cotsmen in membership
...not, however, recognized
...st real international match,
...the 30th of November 1872
...international match between
...yed at Kennington Oval in 1879;
...id and Ireland at Belfast in 1882. In
...national matches were inaugurated with
...stria and Bohemia; and games are now annually
...ith Scotland, Wales, Ireland, France, Belgium, Germany,
...Holland, Austria and other continental countries. As the out-
... to come of the international relations with Scotland, Wales and
When Ireland, an International Football Association Board was formed
by one in 1882, when a universal code of laws was agreed upon. Two
of play representatives f...
ty time stitute the

Martin Boettcher

INVENTOR

only by th... ...rved not
all countries but in
held at Paris o... ...ecting
tion of Association r...tball " was instituted. It consists of the
recognized national associations in the respective countries;
and its objects are to develop and control Association international football. The countries in federation are: Austria,
Belgium, Denmark, England, Finland, France, Germany,
Hungary, Italy, Netherlands, Norway, Sweden and Switzerland.
The small number of clubs taking part in the game in the early
days becomes of interest when compared with the magnitude of

This invention relates to a football kicking tee and more particularly to a tee for holding a football in the desired vertical position by applying a downward pressure on the uppermost end of the ball.

Football kicking tees previously devised for the purpose of holding a ball for practice in kicking extra points or field goals usually only provide a supporting stand or block which position the ball vertically in an appropriately shaped socket or frame. The objective of utilizing a tee is to eliminate the necessity of another player to manually hold the ball during practice sessions as is required during regulation play. Previously devised tees have not proven adequate or satisfactory for this purpose as they are not capable of simulating the manual holding of the ball where a player upon receiving the ball from the center positions the ball in the desired vertical position for the kicker by placing his fingers on the upper end of the ball and applying a slight downward pressure while providing the necessary lateral stability. Such tees have proven inadequate in this respect as they do not simulate the conditions encountered during regulation play. To derive the maximum value from a tee utilized during practice sessions, it is necessary that the tee be capable of applying the slight downward pressure as well as providing the necessary lateral support to more closely simulate a manually held ball.

It is the primary object of this invention to provide a football kicking tee with a vertically spaced holding arm capable of applying the necessary downward pressure on a vertically positioned football in addition to the lateral support thereby closely approximating the conditions encountered when the ball is manually held.

Inventor: James, Elsea Asa
Application Number: US10955461A
Publication Date: 10/01/1963
Filing Date: 05/12/1961

FIG. 2

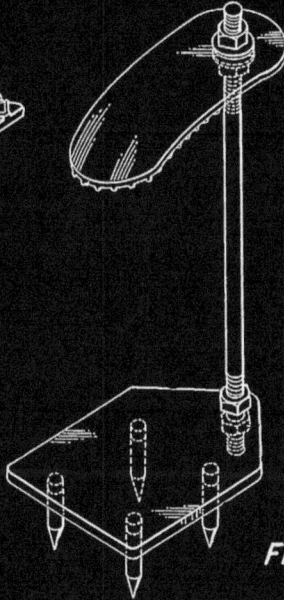

FIG. I

FIG.3

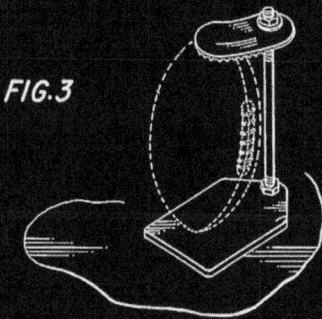

INVENTOR.
ASA J. ELSEA

61

Accurate punt kicking requires for its proper execution such absolute control of the ball that when the player drops it toward his kicking foot the ball, at the instant of impact with the instep of the foot, is properly oriented in relation to the longitudinal axis of the kicking leg and the vertical plane of its upward swing. The football, however, as ordinarily provided for use in the game of football, particularly because of its prolate spheroidal shape, is not readily susceptible of being automatically held in a recognizable position so that it may be accurately dropped in the required oriented position relatively to the kicking foot and it is because of this that very few football players have been able to execute consistently accurate punts despite long periods of practice.

I have found that the art of kicking punts, both end-over-end and spiral, may be developed to a high degree of accuracy and control by practice under conditions which enable the practicing kicker to uniformly drop the ball onto the instep of his kicking foot in correct position for execution of the desired punt and thereby acquire the ability to properly execute the kick even during the pressure of actual game in play.

Having in mind the foregoing, it is among the principal objects of the present invention to provide a football for use in practicing correct and accurate execution of end-over-end and spiral punts which is provided with means for insuring that the ball is always held in a predetermined recognizable hand-held position so that when dropped it is correctly oriented relatively to the inspect of the kicking foot at the instant of impact of the foot against the ball.

More specifically, it is an object of the present invention to provide a football with readily interpretable markings thereon which are constantly visible to the eye of the kicker for properly orienting the ball in his hands preliminary to dropping the same toward his kicking foot.

Still another object of the intention is to provide a football with readily visible guide lines thereon which may be selectively employed by the kicker for his guidance in practice the execution not only of straight-away end-over-end punts, but also spiral punts as executed by the right or left-footed kickers, as the case may be.

Inventor: Thomas, Murray Francis
Application Number: US43820965A
Publication Date: 02/27/1968
Filing Date: 03/09/1965

F. T. MURRAY

FOOTBALL INCLUDING BALL-TO-KICKING LEG ORIENTATION MEANS

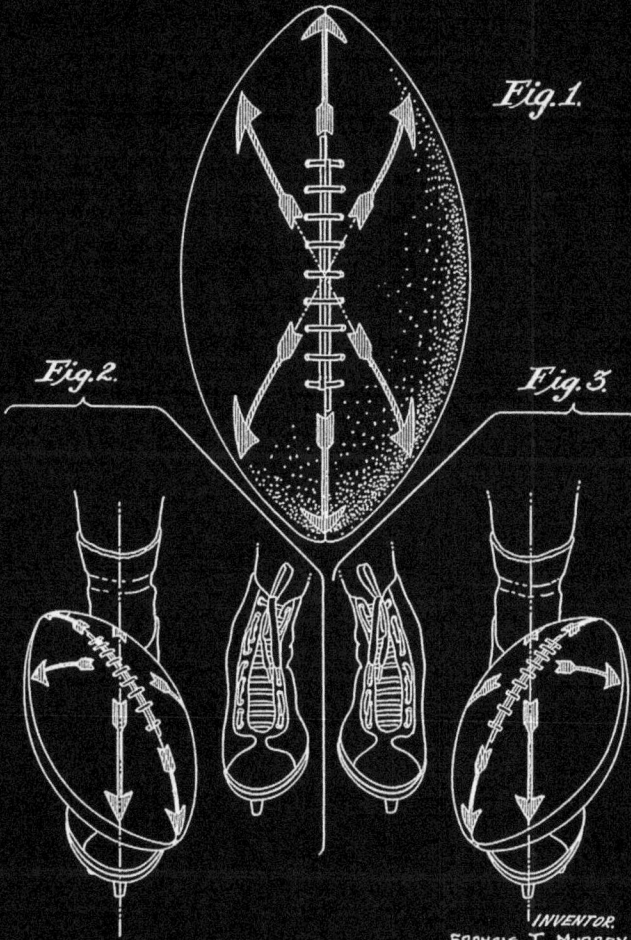

Fig.1.

Fig.2.

Fig.3.

INVENTOR.
FRANCIS T. MURRAY

6. THE FIELD
officiating the game

My invention relates to improvements to football lineman's apparatus, or the like.

The principal object of my invention is to provide apparatus having means of increasing the visibility of measuring units thereof, whereby the referee or umpire, or other official can discern the distance gained, lost, or to be gained after each play more readily than is possible with the type of apparatus at present in common use.

Another more specific object is to provide said means in the form of markers, or targets of such size and color that the units of measurement of the apparatus are clearly distinguishable from a considerable distance.

A further object resides in so supporting the said markers or targets that in case they are made of thin materials, one of the larger surfaces of each will always be in position to be seen from the field.

An additional object is to provide an improved down indicator.

Inventor: Lipp, Julius J.
Application Number: US6629025A
Publication Date: 10/18/1927
Filing Date: 11/02/1925

Oct. 18, 1927.

FOOTBALL LINEMAN'S APPARATUS.

Fig.1

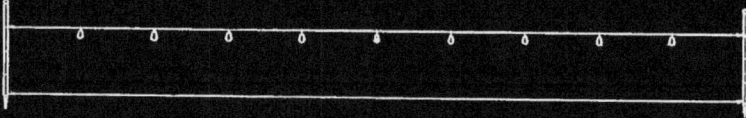

Fig.2

Fig.4

Fig.3

Inventor:
Julius J. Lipp

67

The object of my invention is to devise a novel football score indicator which can be used either as a score card or display device which will visibly indicate the plays as they occur.

It is now customary to broadcast from a radio transmitting station a play by play description of football games, and, due to the broadcasting by the broadcaster of explanatory matter when the game is interrupted, the radio public who are listening in have difficulty in following the different plays and the progress of the game.

My present invention has been especially designed for use, not only by a spectator, but for use by one who is listening to a broadcast description of a game so that, when play is resumed after an interruption, he has a visible record of the score, the approximate location of the ball, the side which has the ball, the number of downs, the period, and the yards to be gained for a first down.

With the above and other objects in view as well hereinafter appear, my invention comprehends a novel football visimeter.

It further comprehends novel means for indicating the approximate position of the ball, the different sides, the possessor of the ball, the score, the period, the number of downs, and the distance which is to be gained.

Inventor: Auer Jr., Gustavus
Application Number: US286663228A
Publication Date: 03/26/1929
Filing Date: 06/19/1928

G. AUER, JR

FOOTBALL VISIMETER

Fig. 1.

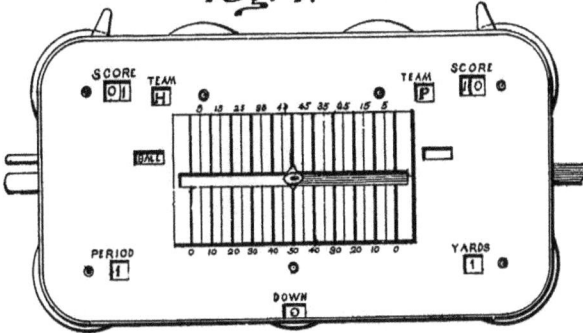

Fig.2.

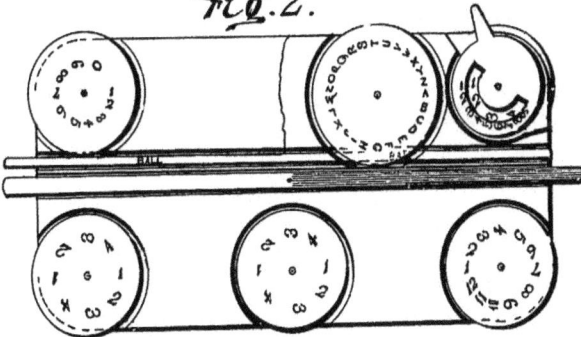

Fig.3.

Fig. 4.

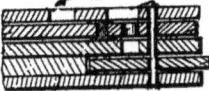

INVENTOR.

Gustavus auer jr

This invention relates generally to football paraphernalia and, more particularly, to a certain new and useful improvement in devices especially adapted for visually designating or indicating the particular "down" in a "four-down gain" period of football of Rugby type.

The game of Rugby football is usually played upon a field of certain dimensions divided between the opposite goals by surface lines or markings into spaces of five yards each, and the particular side or team having the ball must according to the prescribed rules of the game, make a total "gain" toward the goal of the other side or team of ten yards on each of four successive so-called "downs" or suffer the penalty of losing the ball. It is, accordingly, very desirable, if not important, both to the players, officials, and the interested spectators that record be promptly made, and they be informed, of the particular "down" in the so-call four-down period. The object of my present invention is hence to provide, as a new article of manufacture, a simple, inexpensive, and durable device for the purpose, which may be readily portably shifted with and as the players move up or down on the field, which may be conveniently engaged with and on the field in self-supporting upstanding position at the property location, and which may with facility be manipulated to visually indicate both to the players, officials, and spectators, when viewed from either up or down the field, the particular "down" of the set or period.

And with the above and other objects in view, my invention resides in the novel features of form, construction, arrangement, and combination of parts hereinafter described and pointed out in the claim.

Inventor: Schutt, William A.
Application Number: US14245626A
Publication Date: 04/16/1929
Filing Date: 10/18/1926

...idinal lines only and was therefore popularly called th...
"gridiron"; subsequently it was called the "checkerboard."
The end lines are called "goal-lines," the side "touch-lines."
The two lines 25 yds. from each goal-line, and the middle line, ...
...rd-line, are made broader than the rest. In the middle ...
...l-line is a goal, consisting of two uprights exceeding 20 f...
..., set 18 ft. 6 in. apart with a crossbar 10 ft. from th...
...d. The ball is in shape and material of the English Rugby
type.

A match game consists of two periods (*halves*) of thirty-fiv...
minutes with an interval of fifteen minutes. Practice game...
usually have shorter halves. There are four officials: the *umpir*...
whose duty it is to watch the conduct of the players and decid...
regarding fouls; the *referee*, who decides questions regarding th...
progress of the ball and of play; the *field judge* who assist...
the referee and keeps the time; and the *linesman*, who (with tw...

sc...
Casse...
(London,
(" Oval Series
Rugby Football Union Handbook, Richardson, Greenwich, Official
Annual; and *The Football Annual*, Merritt and Hatcher (Association
Game), London.

United States.—In America the game of football has been
elaborated far more than elsewhere, and involves more complica-
tions than in England. From colonial times until 1871 a kind of
football generally resembling the English Association game was
played on the village greens and by the students of colleges and
academies. There was no running with the
ball, but dribbling, called "babying," was
common. In 1871
up, but they wer...
variably ob...
striking ...
allowed
rusher
nutt...
Du...
ba...
co...
th...
Am...
En...
was
in 1...
rules
part
pla...
ad...
r...

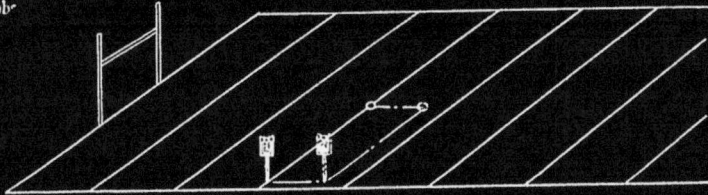

Fig. 1

---- 330 feet ----

DIAGRAM OF FIELD

otball rules provide that when the ball is ...
who receives the ball, commonly known as ...
eyond the line of scrimmage, provided in so ...
m the point where the snapper-back put th...
ward pass may be made provided the ball ...
yds. from the point at which the ball is ...
vals of 5 yds. with white lines parallel ...
uils and for measuring the 10 yds. t...
of 5 yds. with white lines parallel to ...
ermining whether the quarter-back ...
ward pass, such pass is legally mad...
on as in 1902, to what now resen...
exactly how the field should be ...
evenly into 5 yd. spaces, it is wise ...
the field and then to mark off the 5 ...
save labour, it may be sufficient ...
, as the object of these lines is acc...
ansverse lines are distinctly marke...

INVENTOR
William A. Schutt

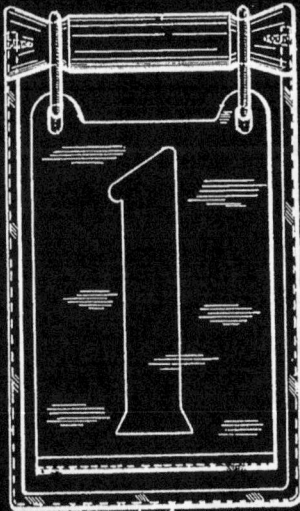

Fig. 2

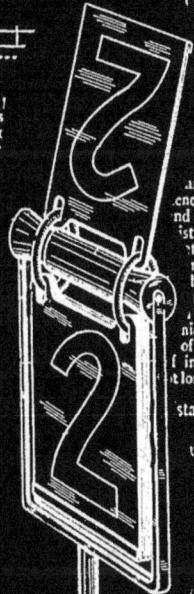

touchdown), a "goal from the
drop-kick, 4, and a "safety"...
2. *Mutatis mutandis*, these
American Rugby differs fr...
scrimmage the men are lin...
separated by the length
man-to-man contest, and
is allowed. Furthermo...
"on side" when ...
passing, *i.e.* thro...
permissible under
consists of a close-fit
and reinforced with ...
and knees, heavy stock...
early period of the game cap...
impossible to keep on, they were ...
wearing of long hair, and the "chrysanthemum head" became
the distinguishing mark of the football player. This, however,

Fig. 3

played ...
ntesting universit.
Later this Rules Co...
ittee of wider repr
well as public and p...
me. The American
September to the ...
rsity of the Americ...
yed in America.
The American Rugby g
a field of 330 ft. long a...
squares with sides 5...
ch side of the field.

ams of eleven men
ided by chalk lines
us as
up the
rom the middle
owing to the
al football is not

As is well understood, in the game of football it is the duty of the linesman, under the direction of the referee, from his position adjacent to one of the side lines of the playing field, to carefully and quickly take note of the varying positions of the ball and whenever a down occurs to properly mark or indicate the distances gained or lost in the progress of the play, such marking or indicating being by the use of a short rod carried by him. Aiding the linesman in the measuring of distances are two assistants who in the performance of their duties use two long rods secured together at their lower ends by a flexible connection ten yards in length.

The utmost accuracy in marking or indicating the position of the ball after each down is, of course, of great importance to both players and spectators, and as indicative of this attention is called to the fact that the rules under which the outstanding games of this country are played provide that the forward point of the ball, in its position when declared dead, shall be taken as the determining point in the measuring. It is therefore frequently a matter of some little difficulty to quickly make the required distance measurement, and especially so when the ball lies near the side line that is farthest from the linesman.

It is the leading object of my invention to provide means that will permit the taking of the required distance measurements more rapidly and accurately than by the present method. Briefly stated I accomplish this object by providing a traveling frame or carriage upon which is located a suitable sighting instrument, such as an ordinary transit, pivotally mounted so as to be swung vertically, and by providing a track for said frame or carriage to travel upon, said track being located adjacent and exactly parallel with one of the side lines of the playing field, and having in connection therewith numbered members indicating distances from the goal lines, with which numbered members suitable indicators or pointers on the frame are adapted to cooperate.

Inventor: Luther, More
Application Number: US27139128A
Publication Date: 06/18/1929
Filing Date: 04/19/1928

June 18, 1929.

L. MORE

MEASURING DEVICE FOR FOOTBALL GAMES

Filed April 19, 1928

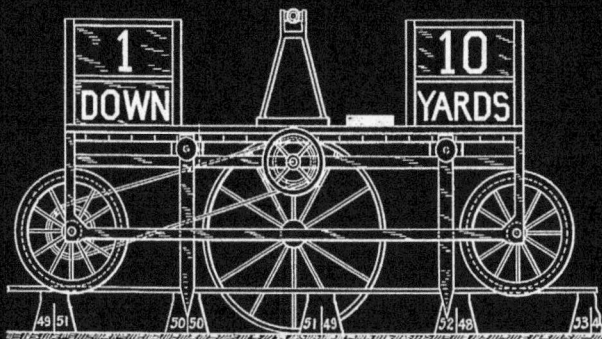

Fig. 1

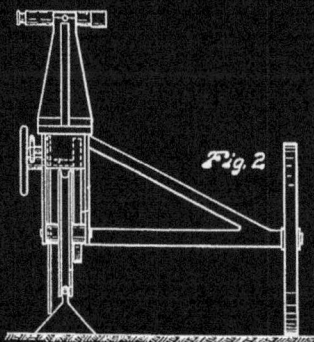

Fig. 2

Fig. 4

Fig. 5

Fig. 6

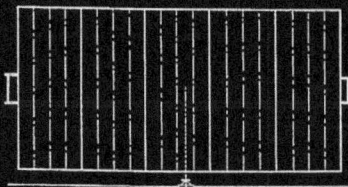

Fig. 3

INVENTOR

Luther More

73

This invention relates to what is believed to be a unique and an effective foul marker for use by officials in football games in satisfactorily designating the area or spot where a foul has occurred, this when the customary signaling horn has been sounded.

Briefly, the marker is characterized by a collapsible fabric base of circular form provided with appropriately located weights, a staff rising vertically and centrally from the base and provided with a foul indicating flag, these parts together with a conical fabric marker attached to the outer perimeter of the base and slidably co-acting with the staff.

The object of the invention is to provide a simple and expedient flag-equipped, weighted marker of a size and type enabling same to be carried in a position in readiness for use in the official's pocket, the structure being foldable and compact for this purpose, but sufficiently weighted to quickly assume a self-standing position when dropped by the official on the spot where the foul ball leaves the playing field.

Inventor: Nicolello, Louis L. D.
Application Number: US56358544A
Publication Date: 10/02/1945
Filing Date: 11/15/1944

Oct. 2, 1945.   L. L. D. NICOLELLO

FOUL MARKER FOR USE BY FOOTBALL OFF

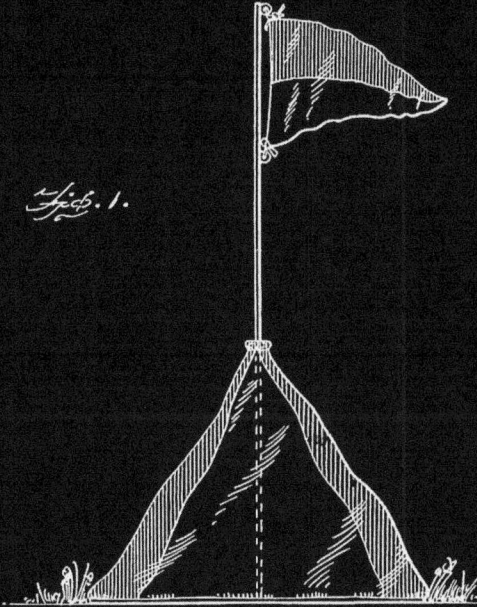

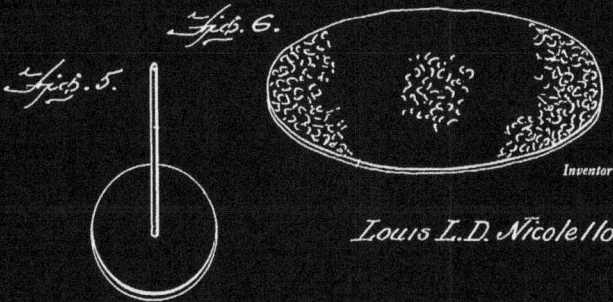

Fig. 1.

Fig. 6.

Fig. 5.

Inventor

Louis L.D. Nicolello

This invention relates to a telescope for use in determining the position of a football longitudinally of a playing field with reference to a previous position of the ball on the field.

As is commonly know, the present method of determining the position of a ball is by means of a chain which for close decisions must be carried out to the location of the ball and referred to the nearest yardage line. This method of determining whether the required yardage has been attained depends upon the accuracy of the yardage line. In many cases, the yardage lines as laid down on the field are not absolutely straight or sometimes become partially obliterated during the course of the game, particularly in inclement weather, resulting in an injustice to one team or the other in the measurement of the yardage.

It is the principal object of this invention to provide a telescope so arranged and mounted that it may be accurately positioned relative to one of the side lines of the field at any selected point on said side line, and sighted across the field at right angles to said side line to determine the position of the ball longitudinally of the field with reference to the line of sight, and therefore with reference to the preselected point on said side line.

A further objection of the invention is to provide a telescope in accordance with the preceding object when the telescope may be swung the horizontal arc of 180 degrees and sighted in opposite directions along said side line and then locked in the position halfway between the diametrically opposed side line sights at right angles to said side line.

Still another object of the invention is to provide a telescope according to any of the preceding objects proved with a target which is plainly visible to the officials on the field.

Inventor: Peresenyi, Louis P.
Application Number: US53814755A
Publication Date: 05/20/1958
Filing Date: 10/03/1955

L. P. PERESENYI

FOOTBALL YARDAGE TELESCOPE

FIG_1

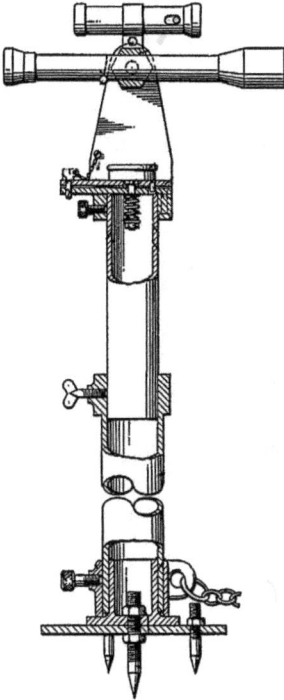

FIG_2

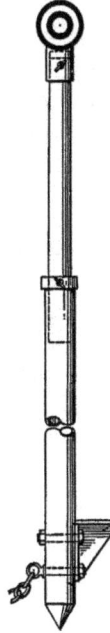

FIG_3

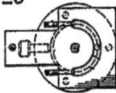

INVENTOR.
LOUIS P. PERESENYI

The game of football is played on a rectangular playing field of greater length than width having goal lines at opposite ends of the playing field, side lines, and having cross lines spaced give or ten yards apart, called yardage lines. On a line and outside of the playing field at each side of the playing field and visible from both sides are numbered discs or markers.

In the playing of the game, the side or team in possession of the football must advance it in four attempts in a forward direction or relinquish the football to the other team, if a ten-yard advance has not been made.

The advance of the football is measured by a lineman with two assistants, sometimes called the chain gang, at a side line of the playing field by two markers ten yards apart and connected by a chain, the rear marker being placed at a side for the first down and other marker being ten yards in advance of the rear marker.

If the question arises as to whether on any down and particularly on fourth down a ten-yard advance has been made, then the linesman assisted by the chain gang carry the two markers to the locality of the football, place the rear marker at a point in lateral alignment with the first down position, and then, assisted by the referee and field judges, measure with the markers and chain to ascertain if a first down has been made. Accurate readings as to ground gained cannot be made in this manner, and honest disputes often occur as to the correctness of the rulings before the final decision can be rendered. Normally five field and quasi-field officials are involved in this operation, while in my invention only the linesman and operator are essential with the referee rendering the decision.

The object, therefore, of my invention is to devise a novel instrument for sighting the exact position of the football on every down by the linesman and his assistants from the side line and provide an accurate measurement for the ground gained in the advance of the football on the playing field.

Inventor: Clime, Henry R.
Application Number: US50722655A
Publication Date: 08/25/1959
Filing Date: 05/10/1955

H. R. CLIME

FOOTBALL SIGHTING DEVICE

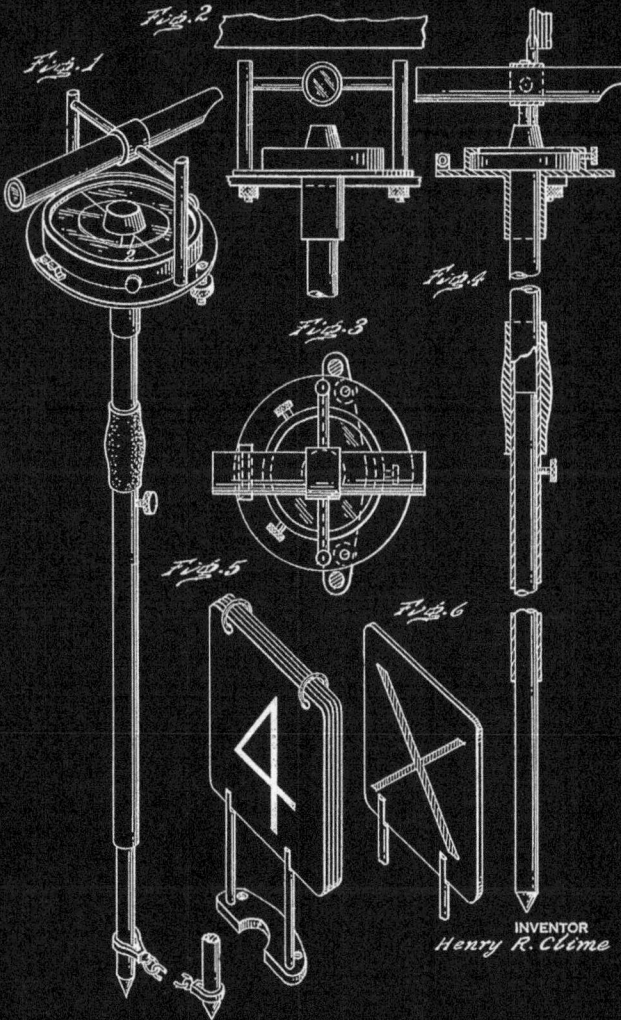

Fig. 1

Fig. 2

Fig. 3

Fig. 4

Fig. 5

Fig. 6

INVENTOR

Henry R. Clime

7. EQUIPMENT
beyond the lines

This invention relates to a football drier, and it concerns more particularly a device for removing moisture from the surfaces of footballs.

During football games played in the rain, or on a wet field, the surface of the ball often becomes wet, and it is customary to wipe the ball with a towel from time to time, or to substitute a new ball, in an effort to maintain the ball in condition for lay.

The covering on a football is ordinarily substantially water repellant, and does not readily absorb moisture, but it is often necessary to subject the ball to a squeezing action to completely free the moisture therefrom. It is an object of the invention to provide apparatus by which the ball is subjected to pressure while rapidly rotating the same whereby all moisture is removed from the cover by centrifugal action, the moisture being converted to a mist which is driven off by a fan.

It is advantageous to both of the opposing teams, in a football game, to keep the ball dry, which makes the ball easier to handle and reduces the possibility of costly fumbles.

An object of this invention is to provide an electrically operable device, for use during football games, adapted to dry footballs quickly while a game is in progress, whereby a dry ball may be kept in play at all times.

Another object of the invention is to provide a device for the purpose described which is of simple, rugged construction, may be manufactured inexpensively, and is efficient in operation and durable in use.

Inventors: Davis Jr., William L. / Baugh, James H.
Application Number: US20276662A
Publication Date: 10/13/1964
Filing Date: 06/15/1962

Dec. 12, 1961

W. L. DAVIS, JR

FOOTBALL DRIER

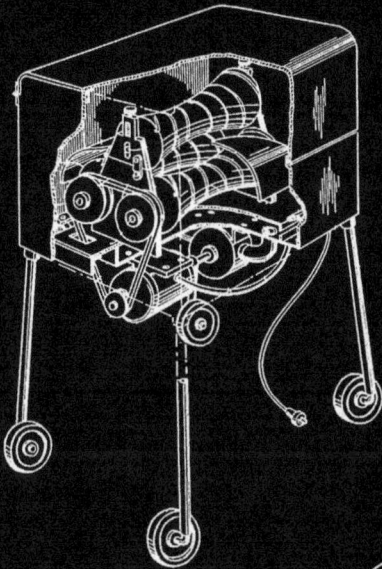

Fig. I

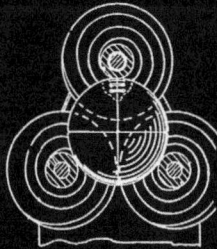

Fig. 3

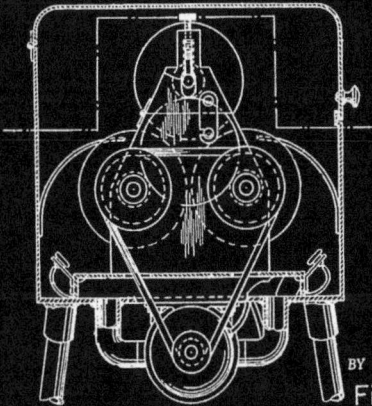

Fig. 2

INVENTOR

William Lynn Davis, Jr.

BY

83

The present invention relates to dryers, and especially to a dryer particularly adapted for rapidly and effectively drying a wet, cold football.

The general object of the present invention is to provide a new and improved dryer particularly characterized by its ability to rapidly and effectively dry and clean footballs by removing water embedded in the pores of the ball, and/or dirt on the surface of the ball.

Another object of the invention is to provide a special receptacle for receiving a support and rotating brush therein in association with a football for knocking off direct particles on the surface of the football and permitting such particles to drop down by gravity through a discharge chute out of the dryer compartment.

Another objet of the invention is to provide a portable, enclosed football dryer unit having a ball receiving bowl or compartment therein in which a ball is received for rotary movement by a support brush in the dryer chamber.

A further object of the invention is to provide a compact, enclosed dyer designed to receive a football therein for rotating and cleaning action thereof while blowing heated dry air into the dryer unit.

Another objet of the invention is to provide a heater and dyer unit particularly adapted for retaining a football, or similar article therein and performing a mechanical abrading and cleaning action thereon while blowing heated air into the football treating unit.

Inventor: Carpenter, Paul O.
Application Number: US83457559A
Publication Date: 02/26/1963
Filing Date: 08/18/1959

Feb. 26, 1963   P. O. CARPENTER

FOOTBALL DRYER

FIG.1

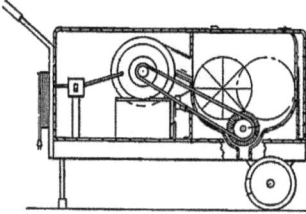

FIG. 2

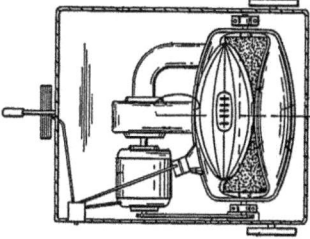

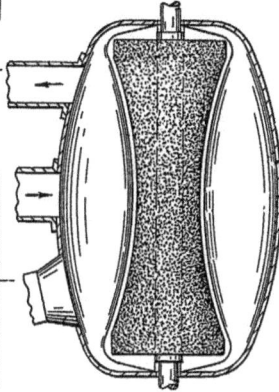

FIG.4

FIG.3

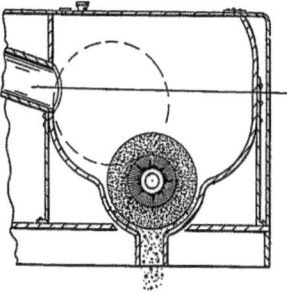

INVENTOR.
PAUL O. CARPENTER

The present invention relates to certain new and useful improvements in a portable apparatus which is expressly and structurally designed and adapted for drying a wet football which has been in play during the course of a football game and which may be, whenever necessary or desired, returned to the field for use.

Briefly summarized the invention is characterized by a suitable mobile manually maneuvered vehicle having a platform. The enclosure or housing is mounted atop the platform and provides a chamber and the driven end of a motor operated shaft projects into the space of the housing and is equipped with clamp means. The clamp means serves, when the cover of the housing is opened, to permit access to be had thereto whereby the ball may be quickly positioned, rotatably spun and the moisture disposed of in part by the centrifugal action, the heated air, which is sucked in at the same time, assisting in the ball drying step.

The invention also features a storage facility in which one or two ready-to-use balls are located in readiness for expedient use.

Inventors: Davis Jr., William L. / Baugh, James H.
Application Number: US20276662A
Publication Date: 10/13/1964
Filing Date: 06/15/1962

Oct. 13, 1964   W. L. DAVIS, JR., ETAL

FOOTBALL DRIER

Fig . 3

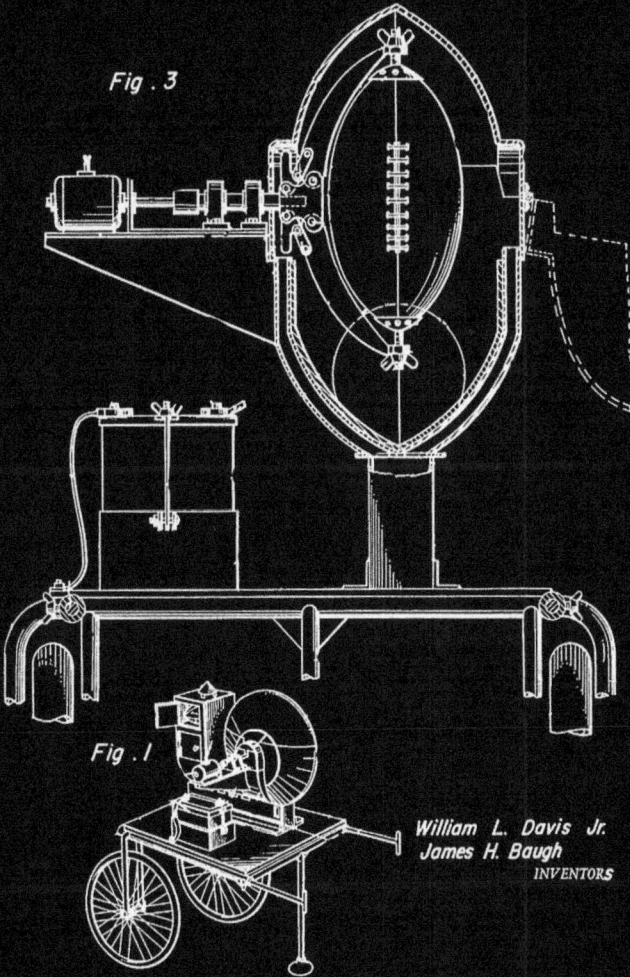

Fig . I

William L. Davis   Jr.
James H. Baugh
        INVENTORS

87

This invention relates to a squad bench designed to prevent injury to a football player who collides with the bench.

A squad bench is almost universally provided to seat football squad members who are not actively participating in a football game. These benches are usually positioned parallel to the side lines of a football field, and only a few yards away from the side lines, to permit quick substitution of players on the bench for active participants.

The traditional football squad bench is constructed of wood and includes a horizontal plank and downwardly projecting legs that support the plank above ground level. Such benches are inexpensive and relatively durable, but constitute an impact hazard to a player whose path during execution of a play intersects the side line. For example, end sweeps and side line pass plays often require both offensive and defensive players to maneuver at top speed in the vicinity of the side line, where their momentum can carry them well outside the side line into the squad bench.

Sharp concerns at the ends of such wooden benches, the projecting legs, and the inelasticity of the wooden benches, make them a potential hazard for players even though they may be wearing the most modern protective equipment.

Accordingly, the present invention provides a squad bench that prevents injury to football players who collide with the bench, and is also strong, durable, and inexpensive to manufacture. The construction of the safety squad bench eliminates sharp projective surfaces, and provides a resilient, impact-absorbing surface on all exposed areas of the bench that may cause injury to a football player.

Inventor: Kahle, Keith Hayes
Application Number: US3506307DA
Publication Date: 04/14/1970
Filing Date: 07/22/1968

April 14, 1970   K. H. KAHLE

FOOTBALL SAFETY BENCH

FIG. 1

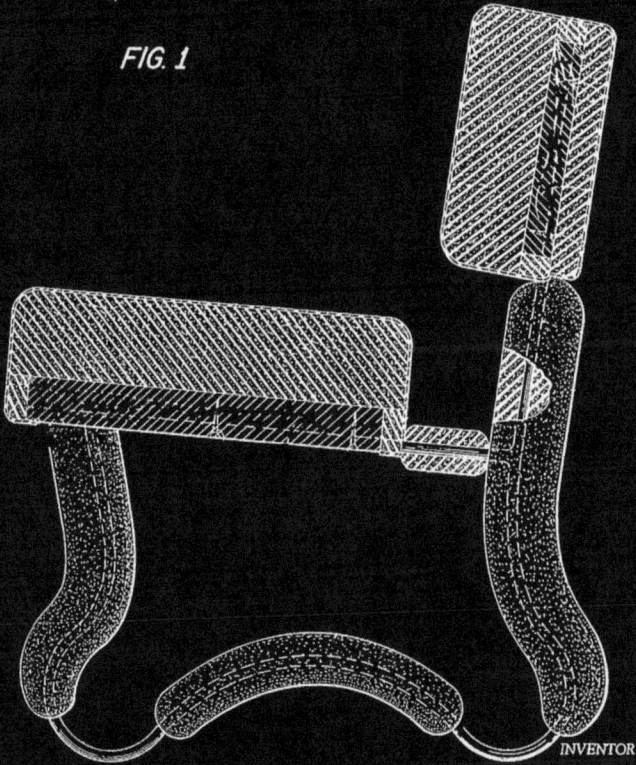

FIG. 2

INVENTOR

KEITH H. KAHLE

89

8. FLAG FOOTBALL
playing safe

This invention relates generally to belts and more particularly to belts for use in playing flag football.

To make football safer to play, various modifications have been made in the game so that the game can be played by youngsters. For example, touch football has been played for many years. More recently, flag football is being played in which the players carry or wear a piece of material, such as a handkerchief carried in a pocket, which can be readily removed by the pursuing player or tackler. However, in playing flag football in this matter it has been found that injuries still have occurred.

In general, it is an object of the present invention to provide a belt which will make it safer to play flag football.

Another object of the invention is to provide a belt of the above character which releasably engages the waist of the player.

Another object of the invention is to provide a belt of the above character in which the belt can be easily grasped.

Another object of the invention is to provide a belt of the above character in which the means for releasably securing the ends of the belt are attached to the belt by particularly novel means.

Inventor: Roselle, Ronald F.
Application Number: US58435056A
Publication Date: 08/05/1958
Filing Date: 05/11/1956

Aug. 5, 1958

R. F. ROSELLE

FLAG FOOTBALL BELT

This invention relates to the playing of football and more particularly to touch football.

Touch football is becoming most popular. However, it is often almost impossible to determine whether or not a touch has been made. Some effort has been made to at least reduce arguments and possibly improper decisions, by the use of handkerchiefs or like tucked under the belt of the players and requiring the removal of same from an opposing player to signify a successful tackle. However, such means for use in playing touch football has many objections. The tightness of the belt, or the manner of placement of the handkerchief or like under the belt may well make it most easy to remove it, or on the other hand, most difficult to remove it. Obviously such non-uniform detachable elements are most unsatisfactory. Also items such as handkerchiefs, towels, or like are not easily seen by either the players or referee.

Therefore, one of the principal objects of my invention is to provide detachable streamers or ribbons for the players of touch football.

More specifically, the object of this invention is to provide a belt and one or more ribbon streamers that has or have an adhesive characteristic that detachably links to the belt.

A further object of this invention is to provide detachable ribbon or streamer for use by players of touch football that is easily seen and interpreted by the players, coaches, referees, and observers.

Still further objects of my invention are to provide a detachable steamer means for players of touch football that is economical in manufacture, durable in use, and refined in appearance.

These and other objects will be apparent to those skilled in the art.

Inventor: Steinkamp, Frederick E.
Application Number: US14424861A
Publication Date: 11/13/1962
Filing Date: 10/10/1961

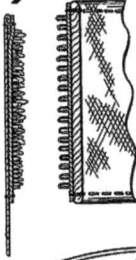

Fig. 1

Fig. 2

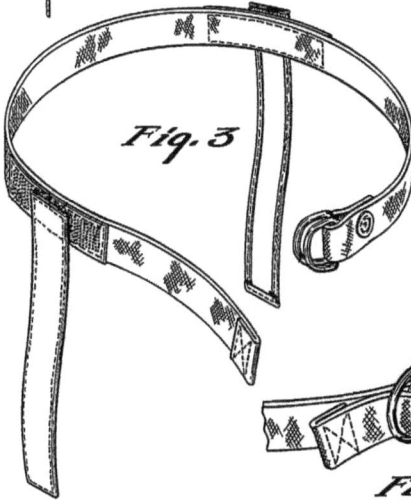

Fig. 3

Fig. 4

95

9. GAMES
just for fun

This invention is a novel game apparatus particularly designed for use in playing imaginary foot ball and for teaching the game of foot ball. The object of the invention is to provide a novel game which will be pleasing to children and adults, and will enable them to play an imaginary game of foot ball understandingly, even if they are not originally familiar with the game, and will not only be interesting as a game in itself, but will also teach the game of foot ball to persons not familiar with the game or the rules thereof.

Inventor: Henry, Wilkins Clyde
Application Number: US46891321A
Publication Date: 04/18/1922
Filing Date: 05/12/1921

99

This invention relates to amusement devices, particularly to games, and has for its object the provision of a game designed to be played upon a playing board for carrying out the action of the ordinary football game, the same requirements applying as in the outdoor game and the same series of plays with the resultant advantage and penalties being also used.

An important object is the provision of a game of this character in which use is made of some counting means such as a pair of differently colored dice, a spinning pointer movable over a graduated dial, or a spinning dish cooperating with a fixed pointer, the number obtained from the use of whatever counting device is used referring to a series of numbers on a chart which bears the names of the various plays together with the advantages and penalties.

Another object is the provision of a game of this character which includes a board which may be subdivided by marks inscribed thereon to define a football playing field or which may be provided a plurality of holes into which may be inserted the markers indicating the location of the ball and the linesman along the field.

Another object is the provision of a game of this character which not only constitutes an amusement device but which is also highly advantageous as means for instructing new players on school teams and the like as to the possible plays and the results thereof.

Inventor: Carl, Head
Application Number: US55191122A
Publication Date: 05/19/1925
Filing Date: 04/12/1922

May 19, 1925.

C. HEAD

INDOOR FOOTBALL GAME

Fig.1.

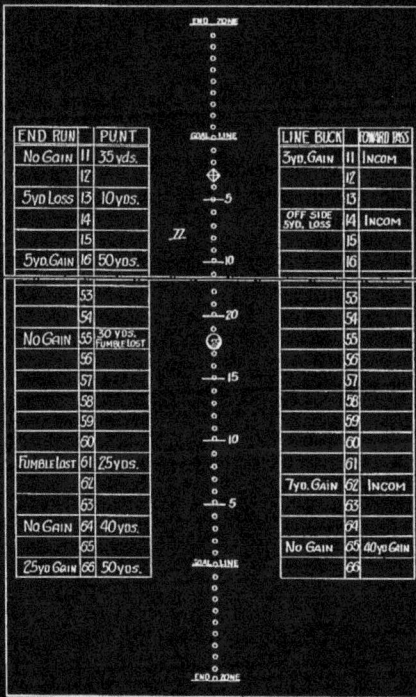

Fig.3.

Carl Head.
INVENTOR

My invention has reference to the playing of table football in which there are provided on a table the usual goal posts and markings to indicate the boundaries of the field. Each player will be provided with the appliance, the subject matter of this invention, for the purpose of playing the ball in imitation of the real game of football.

According to the invention, therefore, I form the figure of a man out of a suitable material, such as wood or sheet metal and support it on the end of a handle or grip. One of the legs of the figure is pivotally mounted on the body and controlled in its movements by a spring cushioned rod in or upon the handle. Thus the pivoted legal may be operated to swing backwards and forwards to kick the ball from goal to goal.

Inventor: Percy, Findlay Algernon
Application Number: US73921224A
Publication Date: 05/26/1925
Filing Date: 09/22/1924

May 26, 1925.

A. P. FINDLAY

DEVICE FOR PLAYING TABLE FOOTBALL

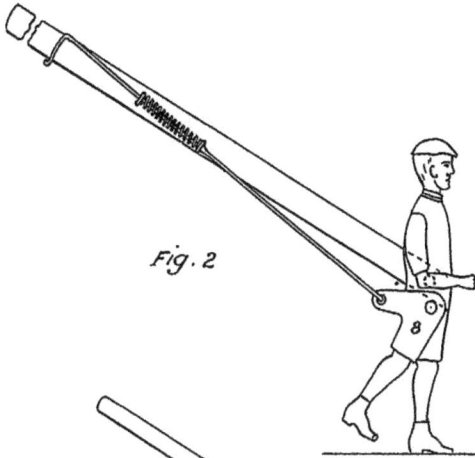

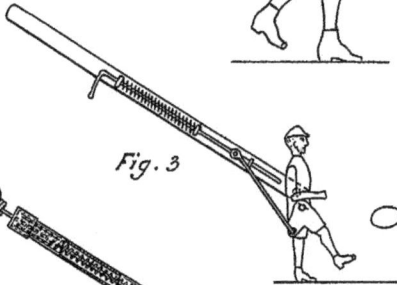

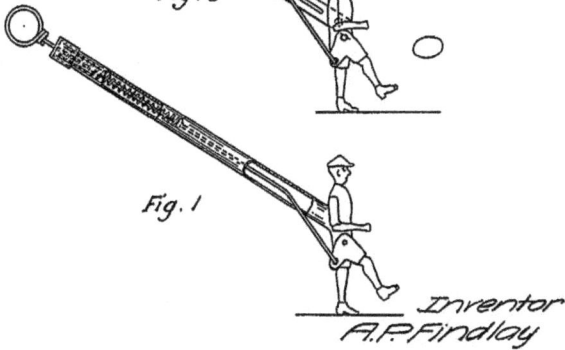

Fig. 2

Fig. 3

Fig. 1

Inventor
A.P. Findlay

103

The primary object of the invention is to provide an improved apparatus suited to playing an imitation game along lines closely related to the plays in actual football games.

A more specific object is to provide an improved chance device for use in determining the successive plays made and the results of such plays.

The invention consists in the novel form, combination and arrangement of parts hereinafter more fully described, shown in the accompanying drawings and claimed.

To begin play the opposing players toss a coin and the winner has the choice of receiving the kick or kicking to his opponent. The player kicking then spins the pointer and the legends in the inner and outer circular areas show the results of the kick-off. For instance, if the inner circle reads "return to 20-yard line".

The player who received the kick is then in possession of the ball on his own 20-yard line and the marker is moved from the kickoff line to his 20-yard line. He then has four downs or spins in which to gain ten yards and failure to make this yardage gives possession to his opponent.

This game is admirably suited for use in teaching persons who are not well versed in football the fundamentals of the game by explaining why certain plays are used under certain conditions, and at the same time the element of chance enters so strongly into it that a beginner often outscores a veteran football player, although superior knowledge of the game is usually shown by the results.

Inventor: Robert, Eugene W.
Application Number: US25673628A
Publication Date: 05/13/1930
Filing Date: 02/24/1928

E. W. ROBERT

FOOTBALL GAME

roved an inadequate protection, and some players now wear a
'head harness" of soft padded leather. Substitutes are allowed
h the places of injured players.

pass or a kick is resorted to, rather than risk losing the ba...
the spot. The kick, although resulting in the loss of the ba...
nevertheless gives it to the enemy much nearer his goal. W...
~ wind is strong the side favoured by it usually kicks o...
other side, not being able to kick back on equal ter...
to play a rushing game, which is always exhaus...
kicking game is often resorted to by the side that ...
the score, in order to save its men and yet retai...
The only remaining way to advance the ball is ...
a free-kick after a fair catch, as in the English game. The ...
kick may be either a punt, a drop-kick or a kick from placem...
Whenever the ball goes over the side line into touch it is bro...
back to the point where it crossed the line by the man ...
carried it over, or, if kicked or knocked over, by a man of ...
side which did not kick it out, and there put in play in on...
two ways. Either it may be touched to the ground and th...

Inventor

Eugene W. Robert,

105

This invention relates to amusement devices, games or the like and more particularly to a game, the success in playing of which is more dependent upon skill than chance.

It is an object of the invention to combine the features of a game requiring skill with the features of a game of popular interest such as the outdoor field game of football.

Another object of the invention is to provide a football "field" for the game having adjustable goals which can be made selectively larger or smaller in area for adapting the device to players of uneven skill so that they will be more evenly matched, thereby providing a greater degree of interest and enjoyment in playing the new game.

A further object of the invention is to provide a game board for playing a game simulating football and upon which the "end runs" of a football game may be simulated and executed.

A still further object of the invention is to provide a game board, the squares of which may be readily indicated for facilitating a playing of the game by mail.

Inventor: Welna, Fred J.
Application Number: US1177735A
Publication Date: 03/30/1937
Filing Date: 03/19/1935

March 30, 1937.   F. J. WELNA

FOOTBALL GAME

Fig.1.

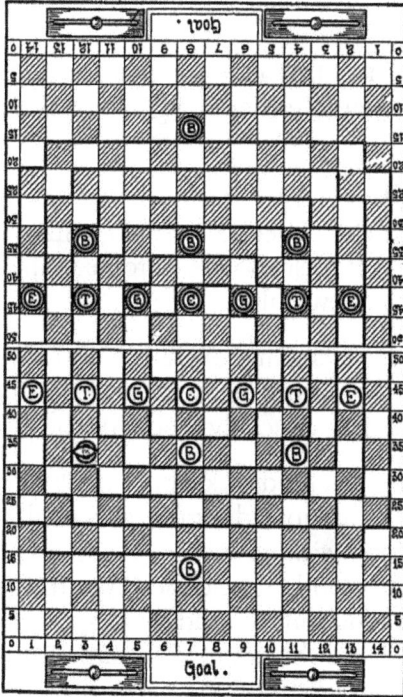

Fig.2.

Fig.3.

Fig.4.

Fig.5.

Fig.6.

Fig.7.

Inventor

F.J.Welna

The present invention relates to a game and to apparatus for playing the same, and particularly to a game simulating the outdoor game of football, and to apparatus for playing such a game indoors.

One object of the invention is to provide a game which two or more persons can readily play with relatively inexpensive equipment and which will engender plays and situations closely resembling those encountered in a regulation game of football, thereby making the game enjoyable not only to regulation football fans but to any players.

Another object of the invention is to provide a game, simulating football, which can be played either as a game of skill or as a game of chance.

Still another object of the invention is to provide a game board for a football game, which is lined as to yardage like a football field, but which is laid out in such wise as to keep the playing space within easy reach of two players.

The game may be played as a game of chance with a spinner or with dice, or as a game of skill, either with a standard deck of playing cards or with a deck specially made for the purpose.

A simulated football field is used to keeping track of play. This field is marked off in graduations designated as yards, and it is one hundred yards long with an ordinary football field. It differs from the standard football field, however, in that it is laid out in a circle. The graduations around one half of the circle are numbered from 1 to 50 and the graduations around the other half of the circle are numbered from 50 to 1. Thus, one half of the circular field corresponds to one half of the playing field of a standard football gridiron and the other half of the circular field corresponds to the other half of a standard football gridiron. With this arrangement, both goal lines and both ends zones are at the same end of the field.

Inventor: Graves, Frank R.
Application Number: US75885847A
Publication Date: 08/16/1949
Filing Date: 07/03/1947

Aug. 16, 1949. F. R. GRAVES

FOOTBALL GAME

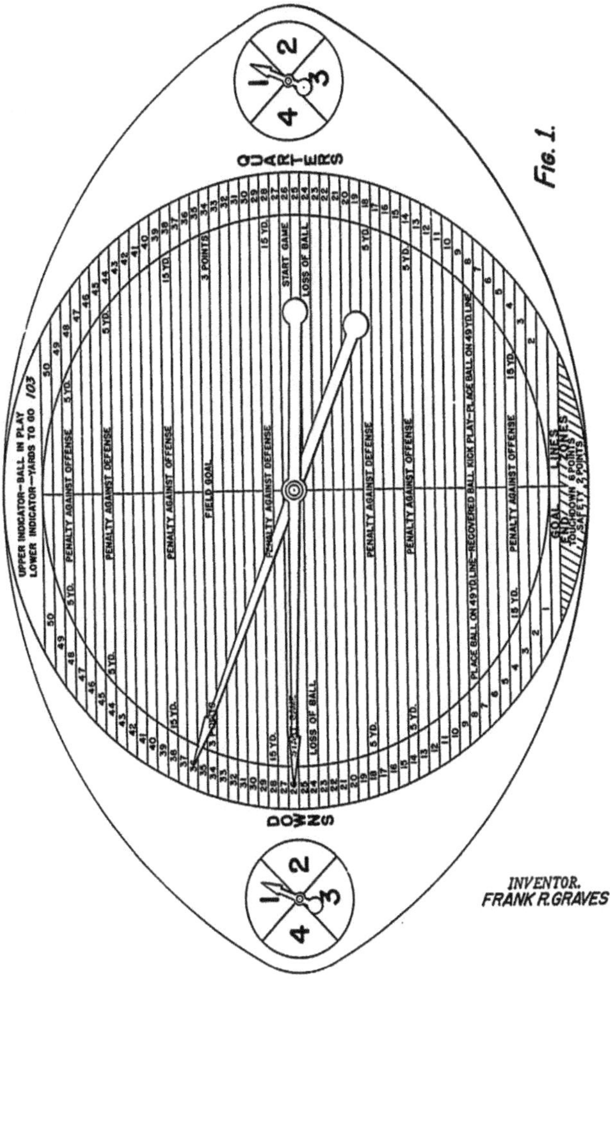

FIG. 1.

INVENTOR.
FRANK R. GRAVES

This invention relates to games wherein opposing teams or players use offensive and defensive maneuvers and wherein a number of types of plays are available to the offense. Under such circumstances the ability of the defense to anticipate which type of play available to the offense will be used enables the former to more effectively block or nullify the play.

An object of the invention is to embody realistically in a game of this kind means whereby actual knowledge of the game and how it should be played may be used to advantage by each player.

Another object of the invention is to eliminate, so far as possible, the factor of chance in directing progress of a game, enabling an alert defense to profit at the expense of the offense, and vice versa.

The invention is well suited to games simulating various types of ports such as hockey, football, baseball, etc., and for purposes of illustration I have selected the game of football.

Inventor: Stringer, Arthur C.
Application Number: US3064348A
Publication Date: 02/26/1952
Filing Date: 06/02/1948

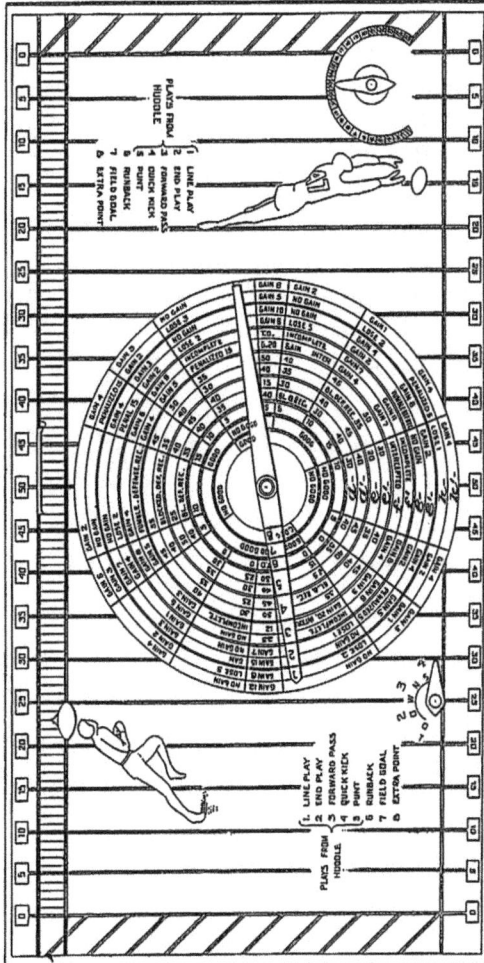

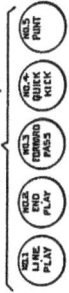

FIG. 1.

FIG. 2.

FIG. 3.

Inventor

A. C. Stringer

111

This invention relates to an indoor or table football game in miniature in which provision is made to simulate the array on an actual playing field and to duplicate the actual physical contacts of the game as determine factors in the progress thereof. In that sense, the game of this invention is to be carefully distinguished from games of the type wherein figures or symbols of players constitute mere markers, the contest proceeding according to the results of separate devices involving chance or skill.

According to the invention, the broad definition of the manner of play involves a catapulting or one or more figures of sensible mass in the general direction of a collection of similar figures representing the opposing team. To the extent that the opposing team may be placed in position, within limits, at the will of one player, and the offensive figures may be directed for catapulting by the other player, the game involves skill and generalship on the part of the players. Upon release of the offensive group, however, the general action proceeds according to the vagaries of chance in a manner suggestive of that involving live players executing individual decisions contributing to a composite result.

The game therefore enables the players to partake of the actual play to the extent of planning the general course of action in each instance and to then relax into the role of a spectator, thus compounding the enjoyment and enlivening the spirit and zest incident to the play.

Inventor: Tullio, John M.
Application Number: US4244948A
Publication Date: 06/17/1952
Filing Date: 08/04/1948

J. M. TULLIO

FOOTBALL GAME DEVICE

FIG.1

FIG. 2

FIG. 3

FIG. 4

FIG. 5

FIG. 6

FIG. 7

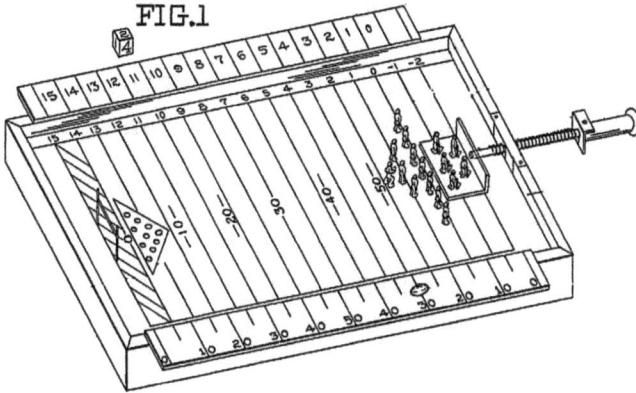

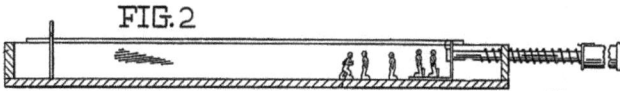

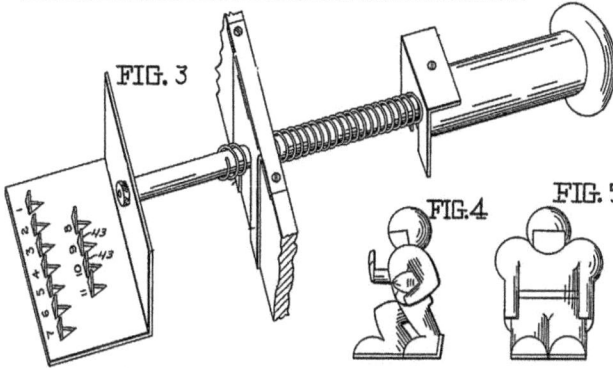

INVENTOR.
John M. Tullio

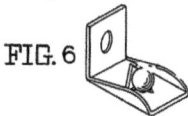

The present invention relates to football players and game and has for an object to provide a game which may employ as few as two players or as many as twenty-two players all of which are under the control of adults or children and which may be played either on a game board or floor.

A further object of the present invention is to provide wheel mounted players which may be propelled either by inertia, spring motor, battery electric motor or manually directed into contact with one another.

A still further object of the present invention is the provision of wheel mounted players which may be remotely controlled.

The basic unit of the game comprises two players, one of which is the ball carrier and other of which is the tackler, the runner being in substantially an erect position and the tackler being in a crouched position with arms extended prior to contact with the ball carrier which arms close about the ball carrier upon the head of the tackler contacting the ball carrier when there is relative motion between the two players. The tackler may actually knock the ball player over.

Inventors: Feather, Franklin G. / Millheim, Stewart B.
Application Number: US54479066A
Publication Date: 04/23/1968
Filing Date: 04/25/1966

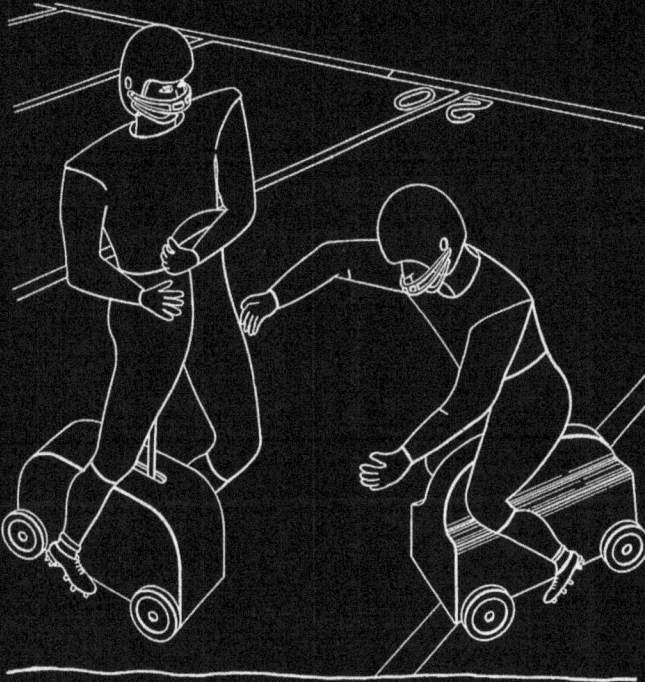

FIG.I.

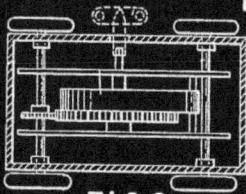

FIG.9.

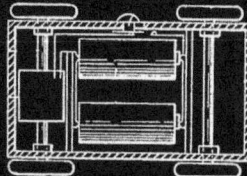

FIG.10.   INVENTORS
Franklin G. Feather &
Stewart B.Millheim

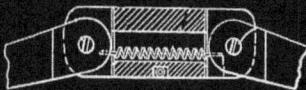

FIG.8.

115

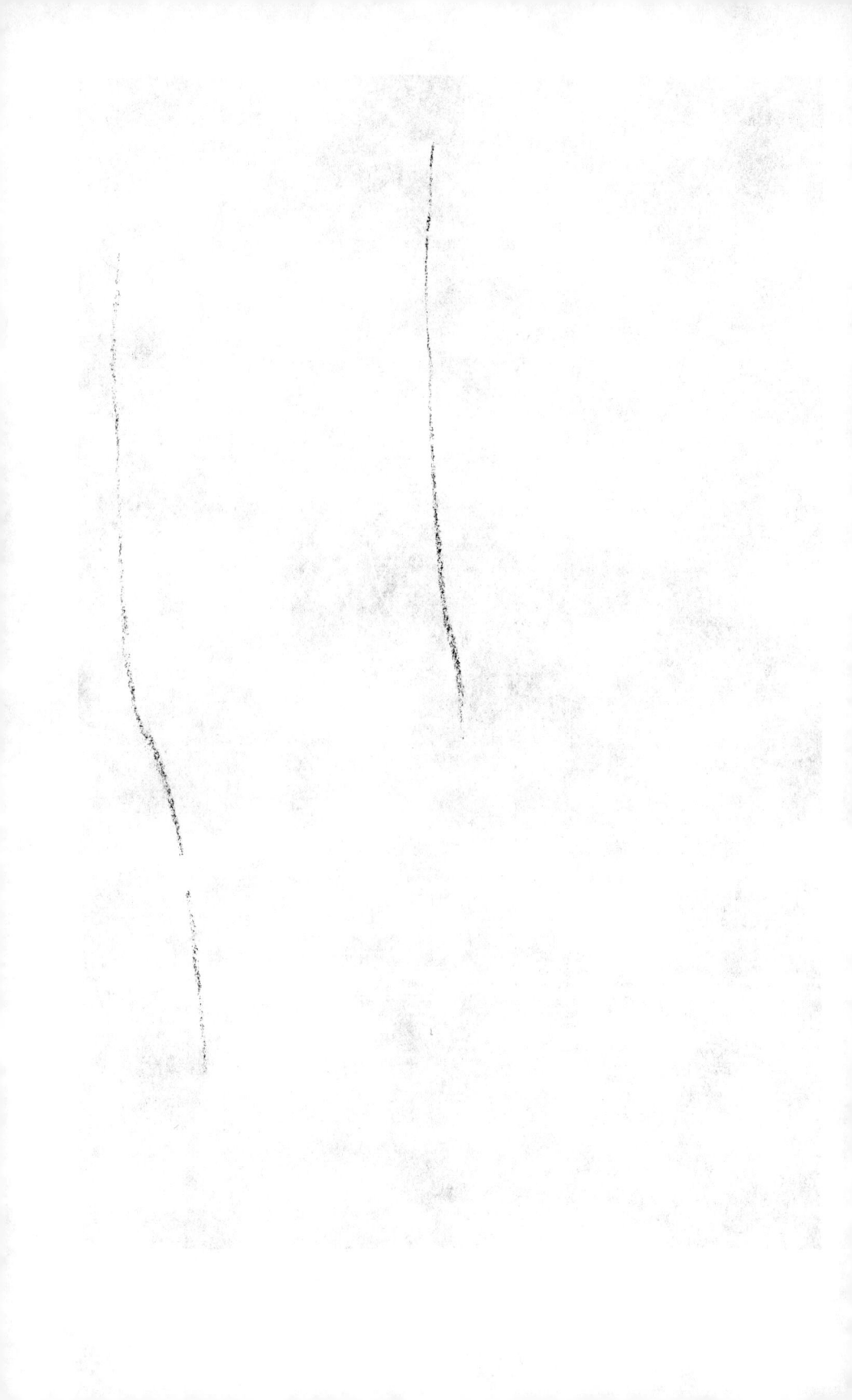

www.ingramcontent.com/pod-product-compliance
Lightning Source LLC
Chambersburg PA
CBHW060621210326
41520CB00010B/1428